Giuseppe Peano

Geometric Calculus
According to the Ausdehnungslehre of H. Grassmann

Translated by Lloyd C. Kannenberg

Birkhäuser
Boston • Basel • Berlin

Lloyd C. Kannenberg, *Translator*
Department of Physics
University of Massachusetts/Lowell
Lowell, MA 01854
U.S.A.

Library of Congress Cataloging-in-Publication Data

A CIP catalogue record for this book is available from the Library of Congress,
Washington D.C., USA.

AMS Subject Classifications: 01A55, 03-03, 15-03

Printed on acid-free paper.
© 2000 Birkhäuser Boston

Birkhäuser

ISBN 0-8176-4126-2   SPIN 10723804
ISBN 3-7643-4126-2

Typeset by TEXniques, Inc., Cambridge, MA.
Printed and bound by Hamilton Printing Company, Rensselaer, NY.
Printed in the United States of America.

9 8 7 6 5 4 3 2 1

Contents

Translator's Note

Traduttore, traditore. A translator's greatest hope is that his work will be *transparent*, presenting no barrier between the author and the reader. A vain hope, alas; yet translations may be useful and even have their virtues. The "King James Bible," for example, is properly regarded as one of the monuments of the English language, yet it is only a translation (and one produced by a government committee at that). I make no special claims for this rendering of Peano's important "take" on Grassmann, which, it appears, was published in a small print run in 1888, and has never been reissued in its entirety (an extract is printed in Peano's *Opera Scelte*, and Hubert Kennedy has included an English translation of the Introduction and Chapter One in his valuable *Selected Works of Giuseppe Peano*), but I do hope it will make the book known to a wider audience.

Of particular interest is, first, Peano's treatment of Grassmann's regressive product, which, although developed only for two and three dimensions, is among the very few discussions of this important operation published until recently, and in fact represents a first step in improving Grassmann's original treatment.* Second, Chapter IX is one of the first attempts to axiomatize the linear vector space idea. In preparing this translation I have taken Kennedy's work very seriously indeed, as will be apparent upon comparing the parts where our work overlaps; but especially in the part on the geometric calculus proper, I have found it impossible to follow his lead in certain renderings. Nevertheless I owe him a great debt, which it is my pleasure to acknowledge here. The patient reader will of course render definitive judgment on the merits, if any, of this effort.

I have silently corrected a few misprints where I discovered them in the text. Otherwise, the layout of the text approximates that of the original.

Finally, I owe special thanks to Alvin Swimmer of Arizona State University for providing me with a photocopy of the original to work from, and to Hongbo Li for reading through the entire manuscript and providing the two Editorial Notes.

---

*See for example, M. Barnabei, A. Brini, and G.-C. Rota, "On the exterior calculus of invariant theory," *Journal of Algebra* **96** (1985), 120–160; A. Brini and A.G.B. Teolis, "Grassmann progressive and regressive products and CG-algebras," in: *Hermann Günther Grassmann (1809–1877): Visionary Mathematician, Scientist and Neohumanist Scholar*, G. Schubring, ed., Kluwer, Boston, 1996, pp. 231–242; A Zaddach, *Grassmanns Algebra in der Geometrie*, Wissenschaftsverlag, Mannheim 1994; A. Zaddach, "Regressive products and Bourbaki," in: *Hermann Günther Grassmann*, pp. 285–295.

I need hardly add that neither they, nor Kennedy, have the slightest responsibility for the inadequacies of this translation; that burden is mine alone.

I am grateful to Ann Kostant and the staff at Birkhäuser Boston for their support, and to Elizabeth Loew for producing a handsome TEX document from my manuscript.

Weston, MA 1997   L. KANNENBERG

Preface

The geometric calculus, in general, consists in a system of operations on geometric entities, and their consequences, analogous to those that algebra has on the numbers. It permits the expression in formulas of the results of geometric constructions, the representation with equations of propositions of geometry, and the substitution of a transformation of equations for a verbal argument. The geometric calculus exhibits analogies with analytic geometry; but it differs from it in that, whereas in analytic geometry the calculations are made on the numbers that determine the geometric entities, in this new science the calculations are made on the geometric entities themselves.

A first attempt at a geometric calculus was due to the great mind of *Leibniz* (1679);[1] in the present century there were proposed and developed various methods of calculation having practical utility, among which deserving special mention are the *barycentric calculus* of *Möbius* (1827),[2] that of the *equipollences* of *Bellavitis* (1832),[3] the *quaternions* of *Hamilton* (1853),[4] and the applications to geometry of the *Ausdehnungslehre* of *Hermann Grassmann* (1844).[5]

Of these various methods, the last cited to a great extent incorporates the others and is superior in its powers of calculation and in the simplicity of its formulas. But the excessively lofty and abstruse contents of the *Ausdehnungslehre* impeded the diffusion of that science; and thus even its applications to geometry are still very little appreciated by mathematicians.

The intent of the present book is the direct explanation, in a form accessible to anyone cognizant of the fundamentals of geometry and algebra, of a geometric calculus, based on some notations contained in the *Ausdehnungslehre*, and in the development of its principal consequences. The geometric entities on which the calculus is performed in this book, are the geometric formations of the 1st, 2nd, 3rd and 4th species (N. 5). The operations which are performed upon them in Chapters 1–4 are:

a) the sum of two formations of the same species (N. 9);

b) the product of a formation with a number (N. 9);

c) the progressive product, or projection, of two formations (N. 10). In Chapters 5–7 there are also introduced:

d) the operations indicated by the symbols $\perp$ and $|$, acting respectively on the vectors of a fixed plane (N. 40) and on vectors and bivectors in space (N. 52);

e) the regressive product, or intersection, of formations (N. 44, 51, 57, 58).

The definitions introduced by means of formations of the fourth species, the volumes, are already common in analytic geometry; the definitions by formations of the first three species are reduced by uniform methods to those given for volumes (N. 6, 7).

The calculus on formations of the first species (points with numerical coefficients and vectors) is identical to the barycentric calculus of Möbius; the reducton of a formation of the first species to a simpler form (Chapter 2) coincides with the determination of the barycenter of several material points, in particular, that of vectors with the composition of translations.

The concepts of line, bivector, formations of the second species, correspond exactly to the expressions of force, couple, system of forces applied to a rigid body of mechanics.

Formations of the third species have no correspondents in mechanics.

The formations of the first species, those of the second that are products of two formations of the first species, and the formations of the third species also correspond, up to a numerical factor, to those in projective geometry called point, right-line {retta}, plane; vectors, bivectors and trivectors correspond to points, right-lines, and planes at infinity.

The geometric formations and the operations on them explained here are all contained in the *Ausdehnungslehre*. But I believe the definitions given, the mode of treatment, and many formulas, are all new.

With Chapter 7, the elementary part of geometric calculus developed in this book may be considered complete. The reader arriving at that point, and taking cognizance of the applications that follow the various chapters, can see all the fecundity and simplicity of this new analysis.

In the succeeding chapters are found applications to the geometric formations of the concepts of infinitesimal calculus, and the enunciation of related theorems, for the most part new.[6] The final chapter treats summarily of the transformation of linear systems in general and of geometric formations in particular. The further development of the questions dealt with in these two chapters I have passed over or deliberately limited.

The geometric calculus is preceded by an introduction that treats of the operations of deductive logic; they present a great analogy with those of algebra and of geometric calculus.

Deductive logic, which forms part of the science of mathematics, has not previously advanced very far, although it was a subject of study by *Leibniz, Hamilton, Cayley, Boole, H. and R. Grassmann, Schröder*, etc.[7] The few questions treated in this introduction already constitute an organic whole, which may serve in much research. Many of the notations introduced are adopted in the geometric calculus.

I will be satisfied with my work in writing this book (which would be the only recompense I could expect), if it serves to disclose among mathematicians some of the ideas of Grassmann. It is however my opinion that, before long, this geometric calculus, or something analogous, will be substituted for the methods actually in use in higher education. It is indeed true that the study of this calculus, as with that of every other science, requires time; but I do not believe that it exceeds that necessary for the study of, e.g., the fundamentals of analytic geometry; and then the student will find himself in possession of a method which comprehends that of analytic geometry as a particular case, but which is much more powerful, and which lends itself in a marvellous way to the study of geometric applications of infinitesimal calculus, of mechanics, and of graphic statics; indeed, some part of such sciences are already observed to have taken possession of that calculus. An assessment of this opinion of mine is the business of the professors, especially those of analytic geometry.

Finally, my thanks to F. Castellano, professor at the Accademia Militare, for volunteering his services in correcting the printing.

Torino, 1 February 1888   G. PEANO

Notes to the Preface

[1] *Leibniz, Math. Schriften*, Berlin 1819, Vol. II., P. 17, and Vol. V, P. 133.
See also, for this purpose, *Grassmann, Geometrische Analyse*, Leipzig, 1847, P. 4.
The propositions that can be expressed with the notations of Leibniz can also easily be expressed with those introduced in the present book.

[2] *Möbius, Der barycentrische Calcul*, Leipzig, 1827.
The notations adopted by Möbius in this and later writings are all reproduced, without alteration, in the present book.

[3] The theory and the more notable applications of the calculus of equipollences are found in the recent work:
Laisant, Théorie et applications des équipollences, Paris, 1887. The fundamental concepts of the calculus of equipollences are also comparable to the calculus that follows, but the notations are somewhat different.

[4] The calculus of quaternions already has a vast literature, which it is useless to cite (*V. Laisant, Introduction à la méthode des quaternions*, Paris, 1881).
The quaternions of Hamilton are not comparable to the calculus developed in the present pamphlet, their being neither geometric formations nor symbols of any of the operations introduced on the formations. An entity that can be considered to have some of the properties of the quaternions is the transformation of vectors in space indicated with the symbol $m + |(\perp^*)$, where m is a number, $\perp$ a vector (N. 86, 2; cf. also N. 83, 14).
The geometric facts that can be expressed with quaternions can also be expressed, and in general with greater simplicity, with the notations of Grassmann developed in the present book.
See E. W. Hyde, "Calculus of direction and position," *American Journal of Math.* **6**(1), where there are collected for comparison some calculations with one and the other method.

[5] The various papers of Grassmann are cited in *Schlegel, System der Raumlehre, nach dem Prinzipien der Grassmann'schen Ausdehnungslehre*, Leipzig, 1872 and 1875.
—, "Grassmann, sein Leben und seine mathematisch-physikalische Arbeiten," *Math. Ann.* **14**(1).
Some of the concepts introduced by Grassmann have been developed, independent of him, by other mathematicians. Cf. besides the preceding article:
Hankel, Vorlesungen über die Complexen Zahlen, Leipzig, 1867, P. 140.
Cayley, "On multiple algebra," *Quart. Journal*, **22** (1867), P. 270.
Other applications of the methods of Grassmann are found in *Caspary, Journal für Math.*, **92**, 123; **95**, 36; **100**, 405; *Math. Ann.* **29**, 581.
Mehmke, Math. Ann. **23**, 143.

He says there: "In the course of some lectures delivered last winter on 'Anwendungen der Grassman'schen Ausdehnungslehre auf Mechanik' I have now convinced myself that the methods of the Ausdehnungslehre can be applied with very great advantage to the definition of centers of gravity, moments of inertia, etc., just as with all other domains of mechanics."

Clifford, American Journal **1**, 350.

Carvallo, Bulletin de la Société math. de France, 1887.

Schendel, Grundzüge der Algebra, nach Grassmann'schen Prinzipien, Halle, 1885.

In addition, to find new applications of this geometric calculus, one ought also to examine the works cited on the calculus of equipollences, and especially those on quaternions.

[6] In my *Applicazioni Geometriche del calcolo infinitesimale*, Torino, 1887, are introduced certain geometric formations whose theory is more amply developed in the present book. In order that the notations adopted in the former be made to coincide with the notations in this one, for the words *segment, area, volume* should be substituted *vector, bivector, trivector*, for the symbol $\equiv$ the symbol $=$, for the expression AB, where A and B are points, $B-A$, and the internal product $U \times V$ of two vectors should be represented by $U \perp V$ or $U|V$, according as it is a region in the plane or in space.

[7] *Leibniz*, op. cit., Vol. V., *Logicae inventionis semina*, and Vol. VII., P. 57.

Cayley, Quart. Journal **11**, 282.

See also, for more extensive bibliographic references: E. Schröder, *Der Operationskreis der Logikkalkuls*, Leipzig, 1877.

In this pamphlet of Schröder (37 pages) is substantially developed the mathematical logic that constitutes the Introduction to the present book. It seemed useful to substitute the symbols $\cap, \cup, -A, \bigcirc, \bullet$ for the logical symbols $x, +, A_i, 0, 1$ used by Schröder, in order to forestall any possible confusion between the symbols of logic and those of mathematics (a thing otherwise advised by Schröder himself). The logical symbols $<$ and $>$ are introduced, which, although not necessary, are very useful (§1, 6). Finally there is introduced the symbol ":" (§4), which, explicitly or implicitly (§5), permits the application of the preceding logical symbols to all propositions, and which is the most important use of the logical symbols introduced; in the geometric calculus that follows is found the grounds for using it in the relations among the propositions. In the paper of Schröder is found incidentally the possibility of applying the logical symbols expressing relations between classes to indicate relations between propositions, but it is not done there; perhaps it will appear in a promised later paper, not published.

See also:

E. Schröder, "Note über den Operationskreiz des Logikkalkuls," *Math. Ann.* **22**, P. 481.

C. S. Peirce, "On the Algebra of Logic," *American Journal* **3**, 15, and **7**, 180.

Clifford, Mathematical Papers, London, 1882.

Jevons, The principles of Science, London, 1883.

Liard, Les philosophes anglais contemporains, Paris, 1878.

The questions relating to mathematical logic are excellent for interesting research. Thus, given n classes, with the logical symbols introduced, it is possible to enunciate $N = 2^{(2^{2n-1})} - 2$ propositions, which can be expressed in functions of 2^n propositions with the operations $\cap, \cup, -$; however one combines with logical operations the N preceding propositions, one always obtains propositions of the same system. For $n = 1$, one has $N = 6$, and, calling A the class, the six propositions are $A = \bigcirc$, $A = \bullet$, $- (A = \bigcirc)$, $- (A = \bullet)$, $- (A = \bigcirc) \cap -(A = \bullet)$, $(A = \bigcirc) \cup (A = \bullet)$. For $n = 2$, $N = 32766$, and so on. There is another question, for which one does not yet see a result: If $a\alpha b$ is a determinate relation between the variable entities a and b, one seeks those classes and the propositions that can be enunciated by means of the relation α, and the logical symbols.

The Operations of Deductive Logic

Notation

§ 1. Let there be a system of entities, and let A, B, ... be classes of this system.

As an example one can consider the system of all real and finite numbers, and as classes of this system, the rational numbers, the integers, the multiples of a given number, the numbers that can be roots of algebraic equations with rational coefficients, and so on. Let a be a number; then one can define the class formed of the numbers greater than a, and which we indicate in this example by $(> a)$, as is likewise defined the class of numbers less than a, which we indicate by $(< a)$.

We introduce the following notation:

1. By the expression $A = B$ we mean the affirmation of the identity of the classes A and B, that is to say that every entity A is also B, and vice-versa.

   The symbol " $=$ " is read *equals*; the proposition $A = B$ is called a *logical equation*; A and B are *members* of this equation.

   Example.
   'even number' $=$ 'multiple of 2'
   'rational number' $=$ 'number developable as a finite continued fraction'

2. By the expression $A \cap B \cap C \cap \ldots$, or $ABC \ldots$, we mean the maximum class contained in the classes A, B, C, ..., that is the class formed of all the entities that are simultaneously A and B and C, etc. The symbol "$\cap$" is read *and*; the operation indicated by the symbol $\cap$ is the *conjunction* of logic; we also call it logical multiplication; the classes A, B, ... are called the *factors* of the *product* $AB \ldots$.

   Example.
   'multiple of 6' $=$ 'multiple of 2' $\cap$ 'multiple of 3'
   '$(> 1) \cap (< 2)$' $=$ 'the system of numbers included between 1 and 2'

3. By the expression $A \cup B \cup C \cup \ldots$ we mean the least class containing the classes A, B, C, ..., in other words the class formed of the entities that are A or B or C, etc. The symbol " $\cup$ is read *or*; the operation indicated by the symbol $\cup$ is

in logic called *disjunction*; we also call it *logical addition*; the classes $A, B, \ldots$ are called *terms* of the *sum $A \cup B \cup \ldots$*.

Example.
  'rational number' = 'integer' $\cup$ 'fraction'
  '(< 1) $\cup$ (> 2)' = 'numbers not included between 1 and 2, and different from 1 and from 2'

4. By the expression $-A$, or $\bar{A}$, we mean the class formed of all the entities not belonging to the class A. The symbol " $-$ " is read *not*; the operation indicated by the symbol $-$ is called *negation*.

Example.
  $-$ 'rational' = 'irrational'
  $-$ ($> a$) = 'class of numbers less than or equal to a'

5. In order that the preceding operations always be meaningful, it is necessary to consider as a class the assembly of all the entities of the system, which is indicated by the symbol ● and is read *all*; it is also necessary to consider as a class the lack of every entity, which is indicated by the symbol ○ and is read *empty*.

Consequently the expression '$A = ○$' represents the proposition 'there are no members of A;' '$AB = ○$' expresses the universal negative proposition 'no A is B; '$A \cup B = ●$ says 'everything is either A or B.'

Example.
  (> 1) $\cup$ (< 2) $= ●$
  (< 1) $\cap$ (> 2) $= ○$
  'integers' $\cup$ 'fractions' $\cup$ 'irrationals' $= ●$
  'rationals' $\cap -$'integers'$\cap$ 'roots of a rational algebraic equation with integer coefficients, and in which the coefficient of the terms of highest degree is unity' $= ○$

6. The expression '$A\bar{B} = ○$' says that there do not exist entities that are A and simultaneously are not B; it is equivalent in substance to the universal positive proposition 'every A is B.' Even though the preceding proposition for indicating that proposition is already quite simple, for greater convenience we will nevertheless also indicate it by the expression

$$A < B \quad \text{or} \quad B > A$$

which can be read 'every A is B' or 'the class B contains the class A.' The symbols "$<$" and "$>$" can also be read *less than* and *greater than*.

Example.

'integers' < 'rationals'   'rationals' > 'fractions'

$(< 1) < (< 2)$ $(> 1) > (> 2)$

Identities

§ 2.   The following identities are obvious:

(1) $AB = BA$ (1′) $A \cup B = B \cup A$

(2) $A(BC) = ABC$ (2′) $A \cup (B \cup C) = A \cup B \cup C$

(3) $AA = A$ (3′) $A \cup A = A$

(4) $A(B \cup C) = AB \cup AC$ (4′) $A \cup BC = (A \cup B)(A \cup C)$

(5) $A \bullet = A$ (5′) $A \cup \bigcirc = A$

(6) $A\bigcirc = \bigcirc$ (6′) $A \cup \bullet = \bullet$

(7) $-(-A) = A$

(8) $- (AB) = (-A) \cup (-B)$ (8′) $- (A \cup B) = (-A) \cap (-B)$

(9) $A \cap -A = O$ (9′) $A \cup -A = \bullet$

(10) $-\bigcirc = \bullet$ (10′) $-\bullet = \bigcirc$

The identities (1), (1′), (2), (2′), and (4) express the commutative and associative properties of the operations ∩ and ∪, and the distributive property of the operation ∩ with respect to ∪. Those properties provide sufficient reason for the names multiplication and addition given to these operations. Formulas (5) and (5′) say that ○ is the identity of the operation ∪, and ● that of the operation ∩. The properties expressed in the identities (3) and (3′) have no correspondents in algebra. Formula (4′), which is little used, expresses the distributive property of the operation ∪ with respect to ∩, in order that the operations ∪ and ∩ be commutative, associative and each distributive with respect to the other; they are the only operations in operational calculus now known to enjoy these properties. The operations ∩ and ∪ do not admit inverse operations in the strict sense of the word; in a certain sense the operation − supplies that lack; formula (7) says that two negations produce an affirmation; (8) says that the negative of a product is the sum of the negatives of the factors, a property analogous to that of logarithms; (8′) says that the negative of a sum is the product of the negatives of its terms. Formula (8′) can also be written $A \cup B = - (-A)(-B)$; thus, introducing the symbols ∩ and − makes it possible to do without the symbol ∪. Formulas (10) and (10′) say that it suffices to introduce only one of the symbols ○ and ●.

The preceding formulas are all obvious, and in every argument they are used continually. Some among them must be taken as axioms, the others as consequences of these; and in fact the remaining formulas can be recovered from (1), (2), (3), (4), (5), (7), (8), (9), (10), by substituting into them, for certain expressions, others that they equal by virtue of these same formulas.

Thus, to recover (8′), one has the series of equations

$$(-A) \cap (-B) \underset{(7)}{=} \{-[(-A) \cap (-B)]\} \underset{(8)}{=} -[-[(--A) \cup (--B)] \underset{(7)}{=} -(A \cup B),$$

by virtue of the identities indicated by the numbers placed under the = symbols; equating the first of the members with the last, the formula is proved.

We also have

$$A \, \bigcirc \underset{(9)}{=} A(A \cap -A) \underset{(2)}{=} A \cap A \cap -A \underset{(3)}{=} A \cap -A \underset{(9)}{=} \bigcirc$$

and equating the first of the members with the last, formula (6) is proved.

The formulas with accents are then deduced in that sense, changing $A, B, \ldots$ into $\bar{A}, \bar{B}, \ldots$ and taking the negative of both members.

There are also the identities

(11) $A \cup AB = A$ 	(11′) $A(A \cup B) = A$

Indeed one has

$$A \cup AB \underset{(6)}{=} A \bullet \cup AB \underset{(4)}{=} A(\bullet \cup B) \underset{(6')}{=} A \bullet \underset{(5)}{=} A$$

and equating the first of the members with the last, formula (11) is proved. Analogously one proves (11′).

Since $A < B$ is identical to $A\,\bar{B} = \bigcirc$, the formulas

(12) $AB < A$ 	(12′) $A \cup B > A$

(13) $\bigcirc < A$ 	(13′) $\bullet > A$

are nothing but new ways of writing the identities

$$AB\bar{A} = \bigcirc, \quad A\bar{A}\bar{B} = \bigcirc, \quad \bigcirc\bar{A} = \bigcirc, A\bar{\bullet} = \bigcirc.$$

§ 3.   An expression obtained by operating on a class X, and on other classes taken to be fixed, with the logical symbols $\cap$, $\cup$, $-$, is said to be a function of X, and is indicated by $f(X)$. By virtue of the preceding identities it may be put into many different forms.

We shall say that a logical expression as a function of X is in a *separated* form if it has the form

(a) 	$$f(X) = PX \cup Q\bar{X},$$

where P and Q are classes independent of X. Every logical function of X can always be put into a separated form, and in one way only. In fact, any class A

independent of X can be put into the separated form

$$A = AX \cup A\bar{X}.$$

The class X can be put into the separated form

$$X = \bullet X \cup \bigcirc\bar{X}.$$

The sum of two separated expressions

$$(PX \cup Q\bar{X})(P'X \cup Q'\bar{X}) = (P \cup P')X \cup (Q \cup Q')\bar{X}$$

remains in a separated form; the product of two separated forms becomes, after some calculation,

$$(PX \cup Q\bar{X}, (P'X \cup Q'\bar{X}) = PP'X \cup QQ'\bar{X},$$

a separated form; the negation of a separated form

$$-(PX \cup Q\bar{X}) = \bar{P}X \cup \bar{Q}\bar{X}$$

also remains expressed in a separated form. Hence every logical function of X that is obtained by operating a finite number of times on constant classes and on X with the operations $\cup, \cap, -$, may be written in a separated form.

It is easy to see the significance of the coefficients P and Q. If in (a) we let $X = \bullet$, and then $X = \bigcirc$, we get

$$f(\bullet) = P, \qquad f(\bigcirc) = Q;$$

hence P and Q have well-determined values. Substituting in (a), we have the relation

(14) $$f(X) = f(\bullet)X \cup f(\bigcirc)\bar{X},$$

which represents an analogy with Taylor's formula.

If $f(X, Y)$ is a logical function of the two variable classes X and Y, and of other classes taken as fixed, we will have, by (14),

$$f(X, Y) = f(\bullet, Y)X \cup f(\bigcirc, Y)\bar{X},$$

and, again applying formula (14) to the coefficients,

(15) $\quad f(X, Y) = f(\bullet, \bullet)XY \cup f(\bullet, \bigcirc)X\bar{Y}, \cup f(\bigcirc, \bullet)\bar{X}Y \cup f(\bigcirc, \bigcirc)\bar{X}\bar{Y}.$

Propositions

§ 4. A proposition may express a relation between determinate entities; it is then said to be *categorical*; considered in itself it can only be true or false. Or a proposition may contain indeterminate (variable) entities, and it is then said to be *conditional*; the condition expressed by such a proposition is in general true of certain entities, and not of others; but the condition could be true of all entities, or of none. Every affirmation is made with categorical propositions; a conditional proposition cannot be part of a categorical proposition. Thus for example the proposition '$x^2 - 3x + 2 = 0$' is conditional, since it contains the indeterminate x; the proposition 'the equation $x^2 - 3x + 2 = 0$ has 1 and 2 as its roots' is categorical, having as subject the preceding conditional. Thus, too, '$(x + y)^2 = x^2 + 2xy + y^2$' is conditional, whereas '$(x + y)^2 = x^2 + 2xy + y^2$ is an identity' is categorical.

If the conditional proposition α contains the indeterminate entity x, by $x : \alpha$ we understand the class formed of all the entities for which the proposition α is true. If a proposition α contains several indeterminate entities $x, y, \ldots$, we consider as a new entity the set of one x, one $y, \ldots$, and we indicate it by $(x, y, \ldots)$. By the expression $(x, y, \ldots) : \alpha$ we understand the class formed by all the entities $(x, y, \ldots)$ for which α is true. If the proposition α does not contain any indeterminate entities other than $x, y, \ldots$, the class $(x, y, \ldots) : \alpha$ is a well-determined class, and a proposition that affirms some property of this class is categorical. If, instead, the proposition α contains, besides the entities $x, y, \ldots$, still other entities $u, v, \ldots$ (and is a function of these entities), the proposition that affirms a property of this class will be a conditional proposition in $u, v, \ldots$

The symbol " :" may be read *such that*.

Example. Let $x, y, \ldots$ be numbers and $f, \phi, \ldots$ be symbols of numerical functions; then:

The expression $x : [f(x) = 0]$ represents the class of numbers x such that $f(x)$ is zero, i.e., the roots of the equation $f(x) = 0$.

The expression $x : [f(x) = 0] \cap x : [f\phi(x) = 0]$ represents the common roots of the two equations $f(x) = 0$ and $\phi(x) = 0$.

The expression $x : [f(x) = 0] = x : [\phi(x) = 0]$ expresses the proposition 'the equations $f(x) = 0$ and $\phi(x) = 0$ have the same roots.'

$x : [f(x)\phi(x) = 0] = x : [f(x) = 0] \cup x : [\phi(x) = 0]$ says that 'the roots of the equation $f(x)\phi(x) = 0$ are roots of $f(x) = 0$ or roots of $\phi(x) = 0$.'

$x : [f(x) = 0] = \bullet$ says that '$f(x) = 0$ is an identity with respect to x.'

$(x, y) : [f(x, y) = 0]$ represents the set of all pairs of values of x and y such that $f(x, y) = 0$.

$x : [f(x, y) = 0] = \bullet$ says that '$f(x, y) = 0$ is an identity with respect to x;' it is a conditional proposition in y.

$y : \{x : [f(x, y) = 0] = \bullet\}$ represents the class of values of y such that $f(x, y) = 0$ is an identity with respect to x.

$y : \{x : [f(x, y) = 0] = \bullet\} = \bigcirc$ says: 'There do not exist values of y such that $f(x, y)$ is zero for every value of x.' This proposition is categorical if $f(x, y)$ is a given function, conditional if $f(x, y)$ is not a given function.

§ 5.   In order to simplify the expression in a discussion when the system of entities $(x, y, \ldots)$ considered as variables is well defined, we shall omit the symbol "$(x, y, \ldots):$" preceding all conditional propositions. We shall adopt this simplification only when all entities represented by letters are considered as variables. Hence, if $\alpha, \beta, \ldots$ are conditional propositions, we have:

$\alpha < \beta$, or $\beta > \alpha$, says that 'the class defined by the condition α is part of that defined by β,' or 'α has β as consequent,' 'β is a consequence of α,' 'if α is true, then β is true.'

$\alpha = \beta$ says that 'if α is true, then β is true, and vice versa.'

$\alpha \cap \beta$ expresses the condition that α and β are true at the same time.

$\alpha \cup \beta$ expresses the condition that α is true or β is true.

$-\alpha$ expresses the condition obtained by negating α.

$\bigcirc$ expresses an *absurd* condition.

$\bullet$ expresses the condition of identity.

If A is a class, $-[A = \bigcirc]$ says 'it is not true that the class A does not contain an entity' or 'the class A contains some entity;' the expression $-[AB = \bigcirc]$ expresses the particular affirmative proposition 'some A is a B.'

Example. In order to acquaint oneself with the symbols introduced, the reader may interpret in ordinary language the following propositions, in which $a, b, \ldots,$ $x, y, \ldots$ represent real finite numbers:

$(a < b) = (b > a), -(a < b) = (a \geq b), -(a = b) = (a \neq b),$

$(a = b) = (a + c = b + c),$

$(a = b) < (ac = bc), (ac = bc) = (a = b) \cup (c = 0),$

$(ac = bc) \cap -(c = 0) < (a = b).$

$(a = b) < (a^2 = b^2), (a^2 = b^2) = (a = b) \cup (a = -b),$

$(ax + b = a'x + b') = [(a - a')x = b' - b],$

$(x + y = a) \cap (x - y = b) = (2x = a + b) \cap (2y = a - b),$

$(xy = 0) = (x = 0) \cup (y = 0),$

$$((x^2 + y^2) = 0) = (x = 0) \cap (y = 0),$$

$$(x^2 - 3x + 2 > 0) = (x < 1) \cup (x > 2),$$

$$-(x^2 - 3x + 2 > 0) = -(x < 1) \cap -(x > 2),$$

$$[(x + y)^2 = x^2 + 2xy + y^2] = \bullet, \ (x^2 + y^2 + 1 = 0) = \bigcirc,$$

$$[x : (ax^2 + bx + c = a'x^2 + b'x + c') = \bullet] = (a = a') \cap (b = b') \cap (c = c'),$$

(i.e., in order for the equality $ax^2 + \ldots = a'x^2 + \ldots$ to be satisfied for every value of x, it is necessary and sufficient that we have at the same time $a = a'$, $b = b'$, $c = c'$),

$$\left\{ x : \left[\frac{1}{x(x-1)} = \frac{a}{x-1} + \frac{b}{x} \right] = \bullet \right\} = (a = 1) \cap (b = -1),$$

$$[x : (x^2 + y^2 = 1) = \bigcirc] = (y < -1) \cup (y > +1)$$

(i.e., in order that the equation $x^2 + y^2 = 1$ not have real roots in x, it is necessary and sufficient that either y be less than -1, or that y be greater than 1).

Operations on propositions

§ 6.   We shall rapidly review some logical identities containing conditional propositions.

Indicating by $A, B, \ldots$ certain classes, or conditional propositions defining these classes, we have the identities:

[1] $\qquad\qquad (A = B) = (B = A),$

[2] $\qquad\qquad (A < B) = (A\bar{B} = \bigcirc),$

[3] $\qquad\qquad (A > B) = (B < A),$

[4] $\qquad (A = B) = (A > B) \cap (A < B).$

The identities [1] and [3] say that: *Every logical expression is changed into an equivalent one by interchanging the two members and changing the symbols $=$, $<$, $>$ into $=$, $>$, $<$.*

The identities [2] and [3] express the definitions given for the symbols $<$ and $>$ by using the symbol $=$; [4] expresses the symbol $=$ using the symbols $>$ and $<$.

Further, we have the following identities, which express the uniformity of the operations $\cap, \cup, -$:

[5] $\qquad\qquad (A = B) < (AC = BC),$

[6] $\qquad (A = B) < (A \cup C = B \cup C),$

[7] $\qquad\qquad (A = B) < (-A = -B).$

Applying the identities [5] and [6] twice we get

[5'] $(A = B) \cap (A' = B') < (AA' = BB')$,

[6''] $(A = B) \cap (A' = B') < (A \cup A' = B \cup B')$.

The identities

[5''] $(A < B) < (AC < BC)$,

[6''] $(A < B) < (A \cup C < B \cup C)$,

[5'''] $(A < B) \cap (A' < B') < (AA' < BB')$,

[6'''] $(A < B) \cap (A' < B') < (A \cup A' < B \cup B')$

may be deduced from the preceding by substituting for the equation $A < B$ its equivalent $A\bar{B} = \bigcirc$. Thus, from [5], we have $(A\bar{B} = \bigcirc) < (A\bar{B}C = \bigcirc)$; now $A\bar{B}C = A\bar{B}C \cup A\bar{C}C = AC(\bar{B} \cup \bar{C}) = AC\bar{B}\bar{C}$ by identities [9],[4],[8]; thus, substituting, $(A\bar{B} = \bigcirc) < (AC\bar{B}\bar{C} = \bigcirc)$ or $(A < B) < (AC < BC)$, which is [5'''].

The preceding identities say that: *From a system of logical equations, all true, and all containing the same sign =, or <, or >, new equations, also true, are deduced by multiplying both members by the same factor or by adding the same term, or summing them member by member, or by multiplying them member by member.*

If in identity [7] we change A and B into $-A$ and $-B$, we deduce that

$$(-A = -B) < (A = B);$$

this identity along with [7] gives, by [4],

[7'] $(A = B) = (-A = -B)$.

We have then

$$(A < B) = (A\bar{B} = \bigcirc) = [\bar{B}(-\bar{A}) = \bigcirc] = [\bar{B} < \bar{A}] = (\bar{A} > \bar{B}),$$

or

[8] $(A < B) = (\bar{A} > \bar{B})$.

Identities [7'] and [8] say that: *Every logical equation is transformed into an equivalent one by taking the negatives of each of its members and changing the symbols =, <, > into =, >, <.*

The following identity is important:

[9] $(A \cup B = \bigcirc) = (A = \bigcirc) \cap (B = \bigcirc)$.

In fact, multiplying both its members by A, we have

$$(A \cup B = \bigcirc) < (A \cup AB = \bigcirc)$$

and, since by (11) $A \cup AB = A$, we have

(α) $\qquad\qquad\qquad\qquad (A \cup B = \bigcirc) < (A = \bigcirc)$.

Multiplying the two members of the proposed equation [9] by B, we obtain analogously

(β) $\qquad\qquad\qquad\qquad (A \cup B = \bigcirc) < (B = \bigcirc)$.

Multiplying (α) and (β) member by member, we deduce that

(γ) $\qquad\qquad\qquad (A \cup B = \bigcirc) < (A = \bigcirc) \cap (B = \bigcirc)$.

Summing $A = \bigcirc$ and $B = \bigcirc$ member by member, we have

(δ) $\qquad\qquad\qquad (A = \bigcirc) \cap (B = \bigcirc) < (A \cup B = \bigcirc)$,

and the combination of (γ) and (δ) says precisely that

$$(A \cup B = \bigcirc) = (A = \bigcirc) \cap (B = \bigcirc).$$

§ 7. By virtue of the preceding identities a logical equation may assume several different forms. Thus the universal affirmative proposition 'every A is B' is indicated by the expression (§1, 6)

(a) $\qquad\qquad\qquad\qquad A < B$.

Exchanging the members [3], it becomes

(b) $\qquad\qquad\qquad\qquad B > A$

or 'the class B contains A.'

Multiplying both members of (a) by A we deduce that

(c) $\qquad\qquad\qquad\qquad A < AB$;

and since we have identity (12) $AB < A$, from this and from (c) we deduce that

(c′) $\qquad\qquad\qquad\qquad A = AB$.

Adding B to both members of (a), we get

(d) $\qquad\qquad\qquad\qquad A \cup B < B$,

which, joined with identity (12′) $A \cup B > B$, gives

(d′) $\qquad\qquad\qquad\qquad A \cup B = B$.

Now multiplying both members of (a) by $\bar{B}$, we will have

(e′) $\qquad\qquad\qquad\qquad A\bar{B} < \bigcirc$

and, by identity (13) $A\bar{B} > \bigcirc$, we have

(e′) $$A\bar{B} = \bigcirc,$$

or 'no A is non-B.'

Adding $\bar{A}$ to both members of (α), we will have

(f) $$\bullet < \bar{A} \cup B$$

and, by identity (13′) $\bullet > \bar{A} \cup B$, we have

(f′) $$\bullet = \bar{A} \cup B,$$

or 'everything is non-A or B.'

Taking the negatives [8] of both members of (a), we have, by exchanging the members,

(g) $$\bar{B} < \bar{A},$$

or 'every non-B is non-A.'

In this way, from proposition (a) we have deduced the others, (b),...,(g), by very simple logical operations. As an exercise, the reader may deduce all the others from another of these propositions. Hence they are all equivalent, and are only different ways of expressing 'every A is B.'

§ 8.   The four propositions

(I) every A is B,

(II) no A is B,

(III) some A is B,

(IV) some A is not B

may be expressed, as we have just seen, by the expressions

(I) $A\bar{B} = \bigcirc$,

(II) $AB = \bigcirc$,

(III) $-(AB = \bigcirc)$,

(IV) $-(A\bar{B} = \bigcirc)$.

Expressions (I) and (II) are called *universal* by logicians; (III) and (IV), which are the negations of universal propositions, are called *particular*; (I) and (III), which contain an even number of negations, are called *affirmative*; (II) and (IV) are *negative*.

We have

$$(AB = O) = (BA = O),$$

or 'no A is B' = 'no B is A.'

We also have

$$-(AB = O) = -(BA = O),$$

or 'some A is B' = 'some B is A.'

These equalities constitute the so-called *inversions* of the universal negative and particular affirmative propositions.

Propositions (I) and (IV), and also (II) and (III), one of which is the negative of the other, are called *contradictory*. We have

$$(AB = O) \cap -(AB = O) = O,$$
$$(A\bar{B} = O) \cap -(A\bar{B} = O) = O,$$

or the coexistence of two contradictory propositions is absurd.

Propositions (I) and (II) are called *contraries*; it is stated in textbooks in logic that two contrary propositions cannot coexist. We have arrived at a somewhat different result. In fact, we have, by formula [9],

$$(AB = O) \cap (A\bar{B} = O) = (AB \cup A\bar{B} = O),$$

or

$$(AB = O) \cap (A\bar{B} = O) = (A = O),$$

that is, the coexistence of propositions (I) and (II) is equivalent to $A = O$; multiplying this equation by $-(A = O)$ we deduce that

[10] $$(AB = O) \cap (A\bar{B} = O) \cap -(A = O) = O,$$

i.e., propositions (I) and (II) cannot coexist, assuming that class A is not empty. Certainly, when the logicians affirm that two contrary propositions cannot coexist, they understand that class A is not empty; but although all the rules given by the preceding formulas are true no matter what the classes that make them up, including O and ●, this is the first case in which it is necessary to suppose that one of the classes considered is not empty.

Formula [10] can also be written

[10'] $$(A\bar{B} = O) \cap -(A = O) < -(BA = O),$$

or 'if every A is B, and if the class A is nonempty, it follows that some B is A.' Hence, given the convention of considering O and ● as classes as well, 'some B is A' is a consequence not of the proposition 'every A is a B' alone, but of this and 'the class A is not empty.'

§ 9. The propositions 'every A is B' and 'every B is C' may be written

$$A\bar{B} = \text{O}, \qquad B\bar{C} = \text{O}.$$

Multiplying the first by $\bar{C}$, the second by A, and summing, we have

$$A\bar{C} = \text{O}$$

or 'every A is C.' Thus we have the simplest form of the syllogism

[11] $(A\bar{B} = \text{O}) \cap (B\bar{C} = \text{O}) < (A\bar{C} = \text{O})$,

which can also be written

[11'] $(A < B) \cap (B < C) < (A < C)$,

or

[11''] $(A\bar{B} = \text{O}) \cap (B\bar{C} = \text{O}) \cap -(A\bar{C} = \text{O}) = \text{O}$,

or, changing C into $\bar{C}$,

[11'''] $(A\bar{B} = \text{O}) \cap -(AC = \text{O}) < (BC = \text{O})$.

This last formula may be stated as: 'If every A is B, and some A, is C, then some B is C.'

If in the forms [11] and [11'''] we change B into $\bar{B}$, or C into $\bar{C}$, or both at the same time, and exchange the factors of the first member, the syllogism takes various forms, some of which have been considered by logicians and called *modes* and *figures*. But among the modes considered by logicians some cannot be reduced to the preceding form, namely those in which from two general propositions is deduced a particular. Thus we cannot obtain the form 'every B is C, and every B is A; therefore some A is C.'

The reason is easy to discover, for while the preceding formulas and the syllogisms derived from them are true whatever the classes introduced, even if these include O or ●, in this new form it is necessary to suppose that the class B is not empty. In fact we have, by formula [9],

$$(B\bar{C} = \text{O}) \cap (B\bar{A} = \text{O}) \cap (AC = \text{O}) = [B(\bar{C} \cup \bar{A}) \cup AC = \text{O}]$$
$$= [B \cap -(AC) \cup AC(B \cup \bar{B}) = \text{O}]$$
$$= \{B \cap [-(AC) \cup AC] \cup AC\bar{B} = \text{O}\}$$
$$= [B \cup AC\bar{B} = \text{O}] = (B = \text{O}) \cap (AC\bar{B} = \text{O}).$$

Hence, multiplying by $-(B = \text{O})$, we have

$$(B\bar{C} = \text{O}) \cap (B\bar{A} = \text{O}) \cap (AC = \text{O}) \cap -(B = \text{O}) = \text{O},$$

which can also be read

$$(B\bar{C} = \bigcirc) \cap (B\bar{A} = \bigcirc) \cap -(B = \bigcirc) < -(AC = \bigcirc),$$

or 'if every B is C, every B is A, and the class B is not empty, some A is C.' In this case, the conclusion therefore depends on three propositions.

§ 10. Finally, we shall treat several questions relating to logical equations. The identities

$$(A < B) = (B > A) = (A\bar{B} = \bigcirc)$$

and

$$(A = B) = (A < B) \cap (A > B) = (A\bar{B} = O) \cap (\bar{A}B = \bigcirc)$$
$$= (A\bar{B} \cup \bar{A}B = \bigcirc)$$

say that each logical equation can be transformed into an equivalent one in which the second member is $\bigcirc$.

The identity

$$(A = \bigcirc) \cap (B = \bigcirc) \cap (C = \bigcirc) \cap \ldots = (A \cup B \cup C \cup \ldots = \bigcirc)$$

says that the system of several coexistent logical equations can be reduced to only one equation, whose second member is the empty set.

An equation or system of equations may contain an unknown class X; we may pose the problem of solving for this unknown. The system of equations having been reduced, for this purpose, to only one whose second member is null, the equation will have the form $f(X) = \bigcirc$, where $f(X)$ represents a logical function of X. By what has been said, $f(X)$ may be put into the separated form $AX \cup B\bar{X}$, and so every equation or system of logical equations can be reduced to the form

$$AX \cup B\bar{X} = \bigcirc$$

or

$$(AX = \bigcirc) \cap (B\bar{X} = \bigcirc),$$

which can also be written

$$(X < \bar{A}) \cap (B < X)$$

or

$$B < X < \bar{A}$$

In order for these equations to be possible, we must have $B < \bar{A}$, or $AB = \bigcirc$. Now, assuming this condition has been verified, it is sufficient to take for X any class whatever containing B and contained in $\bar{A}$, which can be done, if B and $\bar{A}$

are not equal, in an infinite number of ways. The smallest class X is B and the largest is $\bar{A}$; every other can be put in the form $B \cup Z\bar{A}$, where Z is an arbitrary class.

§ 11.   The *elimination* from an equation (or system of equations) of an unknown means writing, where possible, an equation no longer containing the unknown, but containing the other variables in such a way that the proposed equation can be satisfied by any value of the unknown. We have already seen that the result of the elimination of X from the equation $AX \cup B\bar{X} = \bigcirc$ is $AB = \bigcirc$.

The resolution of a system of logical equations with several unknowns can be reduced to the questions already treated by eliminating the unknowns one at a time, as in algebra.

By eliminating a system of variables from one or more equations we get the condition that must hold among the remaining variables in order for the system to be possible. Thus from the equation

$$AXY \cup BX\bar{Y} \cup C\bar{X}Y \cup D\bar{X}\bar{Y} = \bigcirc,$$

eliminating X first we have

$$(AY \cup B\bar{Y}) \cap (CY \cup D\bar{Y}) = \bigcirc$$

or

$$ACY \cup BD\bar{Y} = \bigcirc;$$

eliminating Y from this we have

$$ACBD = \bigcirc$$

as the necessary condition for the proposed equation to be satisfied by certain classes X and Y.

Another application of elimination is the syllogism itself. From two propositions containing three classes (major term, minor term, middle term) we eliminate the middle term in order to have a relation between the other two. Thus, if we want to eliminate B from the propositions $A < B$ and $B < C$, we will have

$$(A < B) \cap (B < C) = (A\bar{B} = \bigcirc) \cap (B\bar{C} = \bigcirc)$$
$$= (A\bar{B} \cap B\bar{C} = \bigcirc).$$

Here the first member is in a separated form, and by eliminating B we have $A\bar{C} = \bigcirc$, which is precisely the conclusion of the syllogism.

On the other hand, from the premises already considered at the end of §9,

$$(B\bar{C} = \bigcirc) \cap (B\bar{A} = \bigcirc),$$

which can be written $B(\bar{C} \cup \bar{A}) = \bigcirc$, and eliminating B we have the identity $\bigcirc = \bigcirc$, thus confirming that from just those two premises no relation between A and C may be concluded.

CHAPTER I

Geometric Formations

Lines, Surfaces, and Volumes

1. Geometric calculus consists in a system of operations analogous to those of algebraic calculus, but in which the entities on which the calculations are carried out, instead of being numbers, are geometric entities which we shall define.

Let $A, B, C, \ldots$ be points in space.

Definition 1. We shall call the right-line {retta} delimited by two points A and B the *line* {linea} AB, and imagine it as described by a point P that moves from A to B.

Given a line AB, where A and B do not coincide, an infinite right-line is determined that contains it, and which we shall call the right-line AB. Its length is also determined; by the expression $mg\, AB$, which is read *magnitude AB*, we shall mean the number that measures the length of this line, with respect to a fixed length, which we shall call the unit of measure. Furthermore, given a line, the direction in which it is imagined to be described is also determined. The lines AB and BA lie on the same right-line and have the same magnitude, but they have opposite directions. Hence they are not to be considered as identical.

Definition 2. We shall call the triangular planar surface described by the line AP, where the point P describes the line BC, from B towards C, the *surface ABC*.

Given a surface ABC, if A, B, C are not on a right-line, the plane that contains it, which we shall call plane ABC, is determined. Its area is also determined, and we shall call the number that measures it, with respect to a fixed area assumed as the unit of measure, $mg\, ABC$.

Given a surface, the sense in which it is imagined to be described is also determined.

Definition 3. We shall call the tetrahedral volume described by the surface ABP, where the point P describes the line CD, in the direction already agreed upon, the *volume $ABCD$*.

By *mg ABCD* we mean the number that measures this volume, with respect to a fixed volume assumed as the unit of measure.

2. Often we shall represent a line, or a surface, or a volume, by a single letter. We shall most often use the letters $a, b, \ldots$ for lines; $\alpha, \beta, \ldots$ for surfaces; and A, B, $\Gamma, \ldots$ for volumes.

If $a, a', \ldots$ represent the lines $PG, P'G', \ldots$ and if α represents the surface XYZ, to the expressions

$$Aa, aA : ABa, AaB, aAB, \alpha A, A\alpha, aa'$$

we shall attribute the meanings, respectively,

$$APQ, PQA; ABPQ, APQB, PQAB, XYZA, AXYZ, PQP'Q'.$$

Operations on volumes

3.   Definition 1. We shall say that a volume A is zero, and write A $= 0$, if its magnitude is zero.

This convention already allows us to express several geometric propositions in a more concise form. Thus:

'$ABCD = 0$' means 'the points $ABCD$ lie in the same plane';
'$A\alpha = 0$' means 'the point A lies in the plane α';
'$ab = 0$' means 'the right-lines a and b lie in the same plane, i.e., either they intersect or they are parallel.'

Definition 2. A volume $ABCD$, which is non-zero, is said to be *right-handed*, or *described in the positive sense*, if a person situated on AB, with head at A and lower extremities at B, sees the surface ABP that describes the volume when the point P moves along the line CD, from C towards D, as moving from left to right. If, instead, the same person sees the plane as moving from right to left, the volume is said to be *left-handed*, or *described in the negative sense*.

Thus, for example, the volume $ABCD$, where B, C, D have the positions shown in the figure, will be right-handed if A is in front of the plane of the page, and left-handed otherwise. The sense of a volume, which is rarely considered in elementary geometry, has great importance in the research we are undertaking.

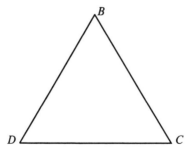

Definition 3. We shall designate as the *ratio of two volumes* A *and* Ω, indicated by $\frac{A}{\Omega}$, the ratio of their magnitudes taken with the $+$ sign or $-$ sign, according as the two volumes have the same sense or opposite senses.

Hence the ratio of two volumes is a real number, which may be positive or negative.

Let Ω be a volume fixed in magnitude and sense. Then:

Definition 4. We shall say that the volumes A and B are equal, and write $A = B$, if

$$\frac{A}{\Omega} = \frac{B}{\Omega}.$$

Definition 5. We shall say that the volume A is equal to the volume B multiplied by the number m, and write $A = mB$, or $A = Bm$, if

$$\frac{A}{\Omega} = m\frac{B}{\Omega}.$$

Definition 6. We shall say that the volume A is the sum of volumes B and Γ, and write $A = B + \Gamma$, if

$$\frac{A}{\Omega} = \frac{B}{\Omega} + \frac{\Gamma}{\Omega}.$$

4. By virtue of these definitions a well-defined meaning has been given to every equality of the form

$$\Sigma = mA + nB + p\Gamma + \dots ,$$

in which m, n, p,... are numbers and A, B, Γ, ... are volumes. It is equivalent to the numerical equality

$$\frac{\Sigma}{\Omega} = m\frac{A}{\Omega} + n\frac{B}{\Omega} + p\frac{\Gamma}{\Omega} + \dots .$$

Hence the same identities hold for volumes as for numbers, and so we have

$$A + B = B + A, \qquad\qquad A + (B + \Gamma) = A + B + \Gamma,$$
$$(m + n)A = mA + nA, \qquad m(A + B) = mA + mB,$$
$$(A = B) < (A + \Gamma = B + \Gamma), \quad (A = B) < (mA = mB),$$
$$(A = B) \cap (B = \Gamma) < (A = \Gamma),$$

which expresses the commutative and associative properties of addition for volumes, the distributive property of the product of a number by a volume, with respect to both factors, and the uniformity of these operations.

We must carefully distinguish the equalities mg A $= mg$ B and A $=$ B; the first says that two volumes have the same magnitude, and the second that they have the same magnitude and sense.

In conformity with algebraic notation, to the expressions $-$A and A $-$ B we shall attribute the meanings (-1)A and A $+ (-1)$B. Hence the equality A $= -$B says that the volumes A and B have the same magnitude but opposite senses.

The inequality $\frac{\pi A}{\pi B} > 0$ means that the volumes πA and πB have the same direction, or that the points A and B lie on the same side of the plane π. Analogously,

$$\left(\frac{\pi A}{\pi B} < 0\right) = \text{(the points } A \text{ and } B \text{ lie on opposite sides of the plane } \pi),$$

$$(\pi A = \pi B) = \text{(the right-line } AB \text{ is parallel to the plane } \pi).$$

Indeed, if $\pi A = \pi B$, the tetrahedrons πA and πB, which have the same base and equal magnitudes, will have the same height; and since they have the same sense, the points A and B lie on the same side of the plane π, and hence the right-line AB is parallel to π; and *vice-versa*.

If in a volume $ABCD$ we interchange two vertices, the magnitude of the volume does not change, but, evidently, the sense changes.

Interchanging one or more vertices, the following identities may be deduced:

$$ABCD = -BACD, \quad ABCD = BCAD,$$
$$ABCD = -BCDA, \quad ABCD = CDAB,$$

which may also be written, on changing letters,

$$ABa = -BAa, \quad AaD = aAD, \quad A\alpha = -\alpha A, \quad ab = ba.$$

Geometric Formations

5.   Definition 1. The set of points A, B, C, ..., to which are affixed, respectively, the numbers m, n, p, ..., is said to be a *formation of the first species*, and is represented by the expression $mA + nB + pC + \ldots$.

Definition 2. The set of lines $a, b, c, \ldots$, to which are affixed the numbers m, n, p, $\ldots$, is said to be a *formation of the second species*, and is represented by the expression $ma + nb + pc + \ldots$.

Definition 3. The set of surfaces $\alpha, \beta, \gamma, \ldots$, to which are affixed the numbers m, n, p, $\ldots$, is said to be a *formation of the third species*, and is represented by the expression $m\alpha + n\beta + p\gamma + \ldots$.

Definition 4. Volumes and sums of volumes of the form $mA + nB + p\Gamma + \ldots$ are said to be *formations of the fourth species*.

The formations of the four species just defined are the geometric entities which comprise the objects of our calculus. For those of the fourth species several operations have already been defined; we shall make the analogous definitions for the first three species depend on those of the fourth in the following way:

6. Definition. A formation of the $\begin{Bmatrix} \text{1st} \\ \text{2nd} \\ \text{3rd} \end{Bmatrix}$ species $\begin{Bmatrix} mA + nB + \ldots \\ ma + nb + \ldots \\ m\alpha + n\beta + \ldots \end{Bmatrix}$ is said to be

zero, and we write $\begin{Bmatrix} mA + nB + \ldots = 0 \\ ma + nb + \ldots = 0 \\ m\alpha + n\beta + \ldots = 0 \end{Bmatrix}$ if, however we take the $\begin{Bmatrix} \text{surface } \pi \\ \text{line } p \\ \text{point } P \end{Bmatrix}$,

we always have $\begin{Bmatrix} mA\pi + nB\pi + \ldots = 0 \\ map + nbp + \ldots = 0 \\ m\alpha P + n\beta P + \ldots = 0 \end{Bmatrix}$.

Hence the equality $ABC = 0$ means that the points ABC have such a position that however we take the point P we have $ABCP = 0$, which evidently means that the area of the triangle ABC is zero. Hence

$$(ABC = 0) = \text{(the points } A, B, C \text{ lie on the same right-line)},$$

$$(Aa = 0) = \text{(the point } A \text{ lies on the right-line } a).$$

The equality $AB = 0$ means that, however we take the points P and Q, we have $ABPQ = 0$, which evidently means that the points A and B coincide. Hence

$$(AB = 0) = \text{(the points } A \text{ and } B \text{ coincide)}.$$

7. Definition. Two formations of the $\begin{Bmatrix} \text{1st} \\ \text{2nd} \\ \text{3rd} \end{Bmatrix}$ species $\begin{Bmatrix} mA + nB + \ldots \\ ma + nb + \ldots \\ m\alpha + n\beta + \ldots \end{Bmatrix}$ and

$$\left\{ \begin{array}{l} m'A' + n'B' + \dots \\ m'a' + n'b' + \dots \\ m'\alpha' + n'\beta' + \dots \end{array} \right\} \text{ are said to be } \textit{equal, } \text{and we write}$$

$$\left\{ \begin{array}{l} mA + nB + \dots = m'A' + n'B' + \dots \\ ma + nb + \dots = m'a' + n'b' + \dots \\ m\alpha + n\beta + \dots = m'\alpha' + n'\beta' + \dots \end{array} \right\}, \text{ if, however we take the } \left\{ \begin{array}{l} \text{surface } \pi \\ \text{line } p \\ \text{point } P \end{array} \right\},$$

$$\text{we always have } \left\{ \begin{array}{l} mA\pi + nB\pi + \dots = m'A'\pi + n'B'\pi + \dots \\ map + nbp + \dots = m'a'p + n'b'p + \dots \\ m\alpha P + n\beta P + \dots = m'\alpha'P + n'\beta'P + \dots \end{array} \right\}.$$

From the preceding definition we deduce:

Theorem 1. *Two formations of the same species that differ only in the order of their terms are equal to each other.*

In fact, letting each term of each formation be followed by a surface π, or by a line p, or by a point P, according to whether it is of the first, second, or third species, we have two sums of volumes that differ only in the order of their terms, and hence are equal. Therefore the proposed formations are also equal.

Theorem 2. *Two formations equal to a third are equal to each other.*

The proof is analogous to the preceding.

8.   The reader may easily prove the following propositions.

Theorem 1. *The equation* $mA = nB$ *says that the points A and B coincide, and that the numbers* m *and* n *are equal.*

Definition. Two lines a and b that lie on the same right-line are said to have the same sense, or opposite senses, according to whether for two points P and Q, arbitrarily chosen, the volumes aPQ and bPQ have the same sense or opposite senses.

Theorem 2. *The equation* $a = b$ *says that the lines a and b lie on the same right-line, have equal magnitudes, and have the same sense.*

Theorem 3. *The equation* $a = mb$ *says that the lines a and b lie on the same right-line, that the ratio of their magnitudes is equal to the absolute value of* m, *and that they have the same sense, or opposite senses, according as* m *is positive or negative.*

Definition. Two surfaces α and β that lie in the same plane are said to have the same sense, or opposite senses, according to whether for an arbitrary point P the volumes aP and bP have the same or opposite senses.

Theorem 4. *The equation* $\alpha = \beta$ *says that the surfaces* α *and* β *lie in the same plane, have equal magnitudes, and have the same sense.*

Corollary. *The equation* $ABC = ABD$ *says that the right-line* CD *is parallel to* AB.

Theorem 5. *The equation* $\alpha = m\beta$ *says that the surfaces* α *and* β *lie in the same plane, that the ratio of their magnitudes is equal to the absolute value of* m, *and that they have the same sense, or opposite senses, according as* m *is positive or negative.*

We also have the identities

$$AB = -BA,$$
$$Aa = +aA.$$

To prove them it is sufficient to follow the members of the first with p, and those of the second with P, thus obtaining the known identities

$$ABp = -BAp, \qquad AaP = aAP.$$

Operations on Formations

9. We now introduce the following operations on formations. For simplicity, by the letters $A, B, C, \ldots$ we mean indifferently points, or lines, or surfaces, or volumes; by the letters $S, S', \ldots$ we mean formations of any species whatever.

Definition 1. We shall designate as the *sum* of two formations of the same species $S = mA + nB + \ldots$ and $S' = m'A' + n'B' + \ldots$, and indicate by $S + S'$ the formation

$$mA + nB + \ldots + m'A' + n'B' + \ldots$$

composed of the terms of the two given formations.

This gives the identities

$$S + S' = S' + S, \qquad S + (S' + S'') = S + S' + S''.$$

Indeed, the two members of these identities contain the same terms.

Definition 2. We shall designate as the *product* of a number x and a formation $S = mA + nB + \ldots$, and indicate by xS, the formation $xmA + xnB + \ldots$ that we obtain by multiplying the coefficient of each of its terms by x.

We have the identities

$$x(S + S') = xS + xS', \qquad (x + y)S = xS + yS'$$

because the members of these identities contain the same terms.

It is quite easy to prove the following propositions:

$$(S = S') < (S + S'' = S' + S''),$$
$$(S = S') \cap (S'' + S''') < (S + S'' = S' + S'''),$$
$$(S = S') < (\mathrm{x}S = \mathrm{x}S'),$$

which say that geometric equations may be summed and multiplied by numbers, just like algebraic equations.

We shall continue to make use of the notations represented by the following equations, in which S, S' are formations and x is a number:

$$S\mathrm{x} = \mathrm{x}S,$$
$$-S = (-1)S,$$
$$S - S' = S + (-1)S,$$
$$\left(\frac{S}{S'} = \mathrm{x}\right) = (S = \mathrm{x}S').$$

The last, which is a logical equation, may be stated: We shall designate as ratio of two formations S and S' that number x which, when multiplied by S, gives S. In order for the ratio $\frac{S}{S'}$ to have meaning, it is first of all necessary for the two formations to be of the same species; but this condition is not sufficient. It results from what has been said that the following have ratios: two volumes, two surfaces that lie in the same plane, and two lines that lie on the same line.

10. The preceding operations are the natural extension of the analogous algebraic operations. We shall now define a geometric operation that has no correspondent in algebra.

Definition. Let $S = m_1 A_1 + m_2 A_2 + \ldots + m_n A_n$ and $S' = m'_1 A'_1 + m'_2 A'_2 + \ldots + m'_n A'_n$ be two formations of species s and s', and suppose that $s + s' \le 4$. Then we shall designate as the *formation projecting from S to S'*, or *progressive product* of the two formations, and indicate by SS', the *formation* $\Sigma_{ij} m_i m'_j A_i A'_j$, i.e., the sum of all the terms obtained from the expression by substituting for i and j all whole numbers from 1 to n and from 1 to n', respectively.

We have, for example:

$$(A + B)C = AC + BC, \quad (B - A)\pi = B\pi - A\pi,$$
$$((B - A)\pi = 0) = (\text{the right-line } AB \text{ is parallel to the plane } \pi),$$
$$((B - A)a = 0) = (\text{the right-line } AB \text{ is parallel to the line } a).$$

The definition of N. 6 can now be stated: A formation of the 1st, or 2nd, or 3rd species is said to be zero if, projecting from it, respectively, a surface, a segment, or a point, arbitrarily chosen, the sum of the volumes so obtained is zero. The definition of N. 7 may also be stated analogously.

We have the following identities:

$$S(S_1 + S_2) = SS_1 + SS_2, \quad (S_1 + S_2)S = S_1S + S_2S,$$
$$S(S_1S_2) = SS_1S_2,$$
$$(xS)S' = S(xS') = x(SS').$$

Indeed the two members of each of these equalities contain the same terms, at most in a different order.

If the formations S and S' are of species s and s', we have

$$SS' = (-1)^{ss'} S'S.$$

Indeed, the two formations SS' and $S'S$ contain the same terms, with the same sign if ss' is even and with opposite signs if ss' is odd.

We now have the following theorem.

Theorem. *If we multiply two geometric equations member by member, we obtain a new equation which is a consequence of those two; i.e.,*

$$(S = S') \cap (S_1 = S_1') < (SS_1 = S'S_1').$$

To make the proof easier, we shall suppose that all formations are of the first species and begin with several particular cases.

(a) $$(S = S') < (SP = S'P).$$

In fact, $(S = S') =$ (whatever the surface π, we have $Sp = S'p) <$ (whatever the line p, we have $SPp = S'Pp) = (SP = S'P)$, which proves (a).

(b) $$(S = S') < (SS_1 = S'S_1).$$

In fact, letting $S_1 = m_1A_1 + n_1B_1 + \ldots$ we have by (a)

$$(S = S') < (SA_1 = S'A_1) < (m_1SA_1 = m_1S'A_1),$$
$$(S = S') < \ldots\ldots\ldots < (n_1SB_1 = n_1S'B_1),$$

$$\ldots\ldots\ldots\ldots\ldots\ldots\ldots\ldots\ldots\ldots\ldots\ldots\ldots$$

Multiplying these logical equations, we have

$$(S = S') < (m_1SA_1 = m_1S'A_1) \cap (m_2SA_2 = m_2S'A_2) \cap \ldots <$$
$$(m_1SA_1 + m_2SA_2 + \ldots = m_1S'A_1 + m_2S'A_2 + \ldots) = (SS_1 = S'S_1),$$

which proves (b).

(c) $$(S_1 = S_1') < (SS_1 = SS_1').$$

In fact

$$(S_1 - S_1') < (S_1 S = S_1' S) = (-SS_1 = -SS_1') = (SS_1 = SS_1').$$

(d) $$(S = S') \cap (S_1 = S_1') < (SS_1 = S'S_1').$$

In fact

$$(S = S') < (SS_1 = S'S_1),$$
$$(S_1 = S_1') < (S'S_1 = S'S_1')$$

and thus

$$(S = S') \cap (S_1 = S_1') < (SS_1 = S'S_1) \cap (S'S_1 = S'S_1')$$
$$< (SS_1 = S'S_1').$$

11.   To summarize what we have said, we have introduced four species of formations, of which particular cases are points, lines, surfaces, and volumes.

We have defined (N. 9) the sum of two formations of the same species, which operation has commutative and associative properties like algebraic addition.

We have defined the product of a number by a formation of any species whatever, and this operation is distributive with respect to both factors.

Finally, we have defined the projection, or progressive product, of two formations of species s and s', assuming $s + s' \leq 4$, which is a new formation of species $s + s'$. This has the distributive property with respect to both factors and is associative; on interchanging the factors the product becomes multiplied by $(-1)^{ss'}$. The product of a number by a formation could be considered as a product of two formations by considering numbers as formations of zero species.

The equality of two formations of the fourth species was made to depend on the equality of two numbers. It was then assumed by definition that two formations of the first, second, or third species are equal if, by following them with the same surface, or the same line, or the same point, the volumes obtained are equal. Finally it was proved that if two formations are equal, then preceding them or following them by the same formation, or by equal formations, yields equal results.

We shall next consider the reduction of formations of successive species to their simplest form, using, for this purpose, a few propositions of geometry and the geometric calculus just explained.

CHAPTER II

Formations of the First Species

Reduction of Formations of the First Species

12.   Theorem. *If A and B are two points,* m *and* n *two numbers whose sum is not zero, then the formation of the first species* mA + nB *is equal to* (m + n)C*, where C is a point of the right-line AB whose distances from the points A and B is inversely proportional to the two numbers* m *and* n*, and which stands between A and B if those two numbers have the same sign, and stand outside the segment AB if they have opposite senses.*

To demonstrate that $mA + nB = (m + n)C$, where A, B, C have the positions indicated, it suffices, by virtue of the definitions given, to demonstrate that, however one takes the triangular surface π, one has $mA\pi + nB\pi = (m + n)C\pi$.

Here is a way to provide that demonstration:

Lemma 1. *If ABC are three points on a straight line, one has $AB + BC + CA = 0$, and vice-versa, that is*

$$(ABC = 0) = (AB + BC + CA = 0).$$

In fact, suppose ABC are on a straight line, and the point B stands in between A and C. If one arbitrarily takes the points P and Q, the volumes $ABPQ$, $BCPQ$ and $ACPQ$ all have the same sense, and the third is the sum of the first two; thus one has, whatever the line PQ:

$$ABPQ + BCPQ = ACPQ$$

which implies that

$$AB + BC = AC$$

which can also be written $AB + BC + CA = 0$. In this demonstration one supposed that B is interposed between A and C; but since that relation is symmetric with respect to A, B, C, it is valid whatever may be the relative positions of those points.

Vice-versa, if $AB + BC + CA = 0$, upon multiplying by C, and observing that $BCC = 0$ and $CAC = 0$, one deduces that $ABC = 0$, QED.

Lemma 2.

$$(ABCD = 0) = (ABC - ABD + ACD - BCD = 0).$$

In fact, if $ABCD$ are situated in a plane, the lines AB and CD either meet in a point X or are parallel. In the first case, since $XAB = 0$ and $XCD = 0$, one deduces from the preceding lemma:

$$AB = XB - XA; CD = XD - XC.$$

Multiply both the members of the first by C and by D, and those of the second by A and by B, one obtains the equations:

$$ABC = XBC - XAC$$
$$ABD = XBD - XAD$$
$$CDA = XDA - XCA$$
$$CDB = XDB - XCB.$$

Summing member by member the first equation with the second changed in sign, with the third, with the fourth changed in sign, one recovers the formula to be demonstrated.

If instead AB and CD are parallel, one has

$$ABC = ABD, \qquad ACD = BCD,$$

from which one again recovers the formula to be demonstrated.

Vice-versa, if $ABC - ABD + ACD - BCD = 0$, multiplying by D one deduces that $ABCD = 0$.

The logical equivalence that constitutes this lemma can also be written

$$(ABCD = 0) = (ABC = D(AB + BC + CA)).$$

This permits us to proceed to the demonstration of the theorem. Since the points ABC are in a straight line, and the segments AC and CB stand inversely as the numbers m and n, one has $mAC = nCB$ or

a) $$mAX + nBC = 0.$$

Let one take at random the triangular surface $\pi = PQR$; the plane π either intersects the right-line ABC in a point X, or it turns out to be parallel.

In the first case, since $XAC = 0$ and $XBC = 0$, one has $AC = XC - XA$, $BC = XC - XB$; thus, substituting into (a), one obtains after some reduction

b) $$(m + n)CX = mAX + nBX.$$

If one multiplies this equality by $PQ + QR + RP$, one has

c)
$$(m+n)CX(PQ+QR+RP) = mAX(PQ+QR+RP)+nBX(PQ+QR+RP),$$

or, since $XPQR = 0$, one has by Lemma 2, $X(PQ + QR + RP) = PQR$, whence

$$(m + n)CPQR = mAPQR + nBPQR,$$

or

d)
$$(m + n)C\pi = mA\pi + nB\pi.$$

Now if the plane π lies parallel to the right line ABC, one has $A\pi = B\pi = C\pi$, from which one recovers relation d). Thus, however one takes the plane π, the result verifies relation d), or

$$(m + n)C = mA + nB.$$

Corollary. *If ABC are in a straight line, and AB are distinct, it is always possible to determine two numbers* m *and* n *such that* $(m + n)C = mA + nB$.

13.   Definition. One calls the *mass* of a formation of first species $m_1A_1 + \ldots + m_nA_n$, where $A_1, \ldots, A_n$ are points, the sum of the coefficients $m_1+m_2+\ldots+m_n$.

Theorem. *Every formation of first species* $m_1A_1 + m_2A_2 + \ldots + m_nA_n$, *whose mass* $m_1+m_2+\ldots+m_n$ *is not zero, is reducible to a single point G with coefficient* $m_1 + m_2 + \ldots + m_n$.

In fact, suppose the number n of the terms of the formation is greater than 2; we can suppose that none of the numerical coefficients are zero, since otherwise the number of terms would be less than n; thus, taking any three terms, e.g., m_1A_1, m_2A_2, m_3A_3, it is not possible for the three equations $m_1 + m_2 = 0$, $m_1 + m_3 = 0$, $m_2 + m_3 = 0$ to hold at the same time; if we suppose, e.g., that $m_1 + m_2 \neq 0$, one can, by the preceding theorem, determine a point G_1, such that

$$m_1A_1 + m_2A_2 = (m_1 + m_2)G_1$$

whence

$$m_1A_1 + \ldots + m_nA_n = (m_1 + m_2)G_1 + \ldots + m_nA_n,$$

that is, the formation proposed, containing n terms, is reduced to another containing $n - 1$ terms, and the mass of the former is equal to that of the latter.

Continuing thus, the formation proposed is reduced to the sum of two terms $pP + qQ$, in which $p + q = m_1 + m_2 + \ldots + m_n$; since this sum is not zero, by hypothesis, one can determine a point G such that $pP + qQ = (p + q)G$, that is

$$m_1 A_1 + \ldots + m_n A_n = (m_1 + \ldots + m_n)G.$$

Note. The point G thus determined is called, in Mechanics, the *barycenter* or *center of gravity* of the points given to which are affixed masses measured by the respective coefficients.

Theorem. *Every formation of the first species whose mass is zero is reducible to the form $P - Q$; one or the other of the two points P and Q can be freely chosen.*

In fact, let the formation $S = m_1 A_1 + \ldots + m_n A_n$, in which $m_1 + \ldots + m_n = 0$. Upon choosing the point Q arbitrarily, one has

$$S = (m_1 A_1 + \ldots + m_n A_n + Q) - Q.$$

But the formation in parentheses, whose sum has the value 1, is reducible to a single point with coefficient 1; one therefore has

$$S = P - Q.$$

14.   Definition. Every formation of the first species of the form $B - A$ is called a *vector*. The points A and B are called respectively the *origin* and *term* of the vector.

Theorem. *In order that two vectors $B - A$ and $B' - A'$ be equal, unless all the points are on one right-line, it is necessary and sufficient that the right-lines AB and $A'B'$ be parallel, as also AA' and BB', or that the figure $ABB'A'$ be a parallelogram.*

In fact, if

a) $$B - A = B' - A',$$

multiplying by $BA'B'$ one deduces, upon neglecting the zero terms, $ABA'B' = 0$, that is the four points $ABA'B'$ lie in a plane.

Multiplying a) by AA' one deduces

$$BAA' = B'AA',$$

that is the right-line BB' is parallel to AA'. Multiplying instead by $A'B'$ one deduces

$$AA'B' = BA'B'$$

that is, the right-line AB is parallel to $A'B'$.

Vice-versa, if the figure $ABB'A'$ is a parallelogram, where O is the point of intersection of the two diagonals, then $2O = A + B'$ and $2O = A' + B$, since O is the midpoint of both diagonals: thus $A + B' = A' + B$, whence $B - A = B' - A'$, QED.

Problems on the Reduction of Formations of the First Species

15. We will solve some problems on the reduction of the formations. These problems are solved by repeating many times the two constructions:

1. To produce the right-line passing through two given points;
2. To produce the right-line passing through a given point and parallel to a given right-line.

These constructions are made with the instruments called *ruler* and *parallels*. For the parallels one could substitute the compass, but we will not introduce it. The problems one can solve by repeating many times the constructions 1 and 2 are called *linear*.

Problem. Given the points A, B, O, to determine the point X for which

$$X - O = B - A.$$

If the points ABO are not on the same right-line, one produces from O the parallel to AB, and from B the parallel to AO; these right-lines intersect in the point X sought.

If instead the point O is on the right-line AB, taking at random the point C outside of the right-line OAB, one constructs D such that $D - C = B - A$, then X such that $X - O = D - C$; then $X - O = B - A$.

Problem. To construct the sum $O + (B - A)$ of a point O and a vector $B - A$.

If one constructs X such that $X - O = B - A$, then

$$X = O + (B - A).$$

Problem. To construct the sum of several vectors $(B_1 - A_1) + \ldots + (B_n - A_n)$.

Taking at random the point O, one determines the points $X_1, \ldots, X_n$ for which

$$X_1 - O = B_1 - A_1, X_2 - X_1 = B_2 - A_2, \ldots, X_n - X_{n-1} = B_n - A_n.$$

Summing, one has

$$X_n - O = (B_1 - A_1) + (B_2 - A_2) + \ldots + (B_n - A_n).$$

Problem. To construct a vector equal to $m(B - A)$, where m is an integer.

If m is a positive integer, the vector sought is the sum of m vectors equal to $B - A$, and one constructs it by the preceding rule.

If m is a negative integer, one puts $m = -m'$, whence $m(B - A) = m'(A - B)$; and thus we reduce it to the preceding case.

Problem. To construct the barycenter C of two given points A and B to which are affixed the integers m and n.

The barycenter should certainly lie on the right-line AB. Taking at random a point O outside of this right-line, one constructs the points A', B' and C' for which

$$A' - O = m(A - O); B' - O = n(B - O),$$
$$C' - O = (A' - O) + (B' - O).$$

One has $C' - O = mA + nB - (m + n)O$, or, since $mA + nB = (m + n)C$, one has

$$(m + n)C = C' + (m + n - 1)O.$$

This formula says that C is on the right-line OC'; thus C is the point of intersection of the right-lines AB and OC'.

Problem. To construct a vector equal to $\frac{m}{n}(B - A)$, where m and n are integers.

One constructs the point C such that

$$nC = (n - m)A + mB;$$

this is the barycenter of the points A and B to which are affixed the integers $n - m$ and m, whose sum one tacitly supposes is not zero. One obtains

$$n(C - A) = m(B - A)$$

or

$$C - A = \frac{m}{n}(B - A).$$

Problem. To reduce to a simpler form any formation of the first species:

$$S = m_1 A_1 + m_2 A_2 + \ldots + m_n A_n,$$

supposing the coefficients are rational.

Taking at random the point O, one has:

$$m_1 A_1 + \ldots + m_n A_n = (m_1 + \ldots + m_n)O + m_1(A_1 - O) + \ldots + m_n(A_n - O).$$

If one puts m $= m_1 + \ldots + m_n$, and calls I the vector to which the sum $m_1(A_1 - O) + \ldots + m_n(A_n - O)$ is reducible, one has

$$S = mO + I;$$

that is, every formation of first species is reducible to an arbitrary point O, to which is affixed the sum of all the coefficients, which is the mass of the formation, plus a vector I.

If m $\neq 0$, setting $G = O + \frac{1}{m}I$, one has

$$S = mG,$$

where G is the barycenter of the given points.

If instead m $= 0$, one has $S = I$, whence the formation is reduced to a vector, which can also be set to zero.

16. Definition. Two vectors $B - A$ and $B' - A'$ are called *parallel*, or *have the same direction*, if the right-lines AB and $A'B'$ are parallel. A vector $B - A$ is called parallel to a right-line P, or to a plane π, if the right-line AB is parallel to P, or to π.

Theorem 1. *If the vectors I and J are parallel, and the first is not zero, one can determine a number* x *such that $J = xI$.*

In fact, taking at random the point A, one sets $B = A + I, C = A + J$; the points A, B, C are on a straight line; and since A and B are distinct, one can determine the numbers m and n for which $(m + n)C = mA + nB$, from which one recovers $(m + n)(C - A) = m(B - A), C - A = \frac{m}{m+n}(B - A)$, and setting $\frac{m}{m+n} = x$, one has $J = xI$.

Theorem 2. *If the vectors I and J are parallel, one has $IJ = 0$, and vice-versa.*

In fact, if I and J are parallel, one determines the number x such that $J = xI$, whence $IJ = xII = 0$.

Vice-versa, taking at random A, and setting $B - A = I, C - A = J$, if $IJ = 0$, one deduces that $(B - A)(C - A) = 0$, whence $A(B - A)(C - A) = 0$, and consequently $ABC = 0$; thus the three points ABC are on a straight line, and the vectors I and J, parallel to the right-line ABC, are parallel to one another.

The propositions

$$((B - A)a = 0) = \text{(the right-line } AB \text{ is parallel to the right-line } a)$$

and

$$((B - A)\pi = 0) = \text{(the right-line } AB \text{ is parallel to the plane } \pi),$$

upon setting $B - A = I$, and making use of the definitions introduced, become

$$(Ia = 0) = \text{(the vector } I \text{ is parallel to the right-line } a).$$
$$(Ip = 0) = \text{(the vector } I \text{ is parallel to the plane } \pi).$$

Theorem 3. *If ABC are three points on a straight line, one has*

(1) $$AB + BC + CA = 0,$$

and vice-versa.

In fact, if ABC are in a straight line, the vectors $B - A$ and $C - A$ are parallel; thus $(B - A)(C - A) = 0$, which, upon development, gives the preceding formula. Vice-versa, if (1) is verified, one deduces $ABC = 0$, and the points A, B, C are in a straight line. This proposition is nothing but Lemma 1 of Theorem N. 12.

Identity (1) can also be written:

(2) $$AB = C(B - A).$$

17.   Theorem 1. *If the vectors I J K are parallel to the same plane, and $IJ \neq 0$, it is possible to determine numbers x and y for which one has*

$$K = xI + yJ.$$

In fact, taking at random the point O, one sets $A - O = I$, $B - O = J$, $C - O = K$; the points $OABC$ lie in the same plane, and the right-lines OA and OB (parallel to I and J) are not parallel to one another. One produces from C the parallel to OB, which intersects OA in A': thus $C - O = (C - A') + (A' - O)$; now $A' - O$, which is parallel to OA, and thus to I, one can put into the form xI; furthermore $C - A'$, which is parallel to J one can put into the form yJ; thus one has $K = xI + yJ$.

Theorem 2. *If I J K are three vectors parallel to the same plane, one has $IJK = 0$, and vice-versa.*

In fact, if IJ is not zero, determining numbers x and y such that $K = xI + yJ$, one deduces $IJK = 0$. If $IJ = 0$, one also has $IJK = 0$.

Vice-versa, if $IJK = 0$, taking at random, as in the preceding theorem, the point O, and setting $A - O = I$, $B - O = J$, $C - O = K$, so that

$$(A - O)(B - O)(C - O) = 0,$$

whence one deduces

$$O(A - O)(B - O)(C - O) = 0,$$

or, developing this, $OABC = 0$; thus the four points $OABC$ lie in a single plane, and the vectors I, J, K, which are parallel to the lines OA, OB, OC lying in the same plane, are parallel to that plane.

Theorem 3. *If $ABCP$ are four points of the same plane, and $ABC \neq 0$, one can determine numbers* xyz *for which one has the result* $P = xA + yB + zC$.

In fact, since the vectors $B - A, C - A, P - A$ are parallel to the same plane, and the first two are not parallel, one can determine numbers m and n in such a way that $P - A = m(B - A) + n(C - A)$, whence one recovers

$$P = (1 + m + n)A + mB + nC,$$

and thus P is expressed in the form required.

Theorem 4. *If $ABCP$ are four points of the same plane, one has*

$$ABC - ABP + ACP - BCP = 0,$$

and vice-versa (N. 12, Lemma 2).

In fact, since the vectors $B - A, C - A, P - A$ are parallel to the same plane, one has $(B - A)(C - A)(P - A) = 0$, which, upon development, gives the proposed identity. One can also write it as

$$ABC = (AB + BC + CA)P.$$

Vice-versa, if this relation is satisfied, multiplying by P one deduces $ABCP = 0$, whence the four points are in a plane.

18.   Theorem 1. *If $IJKU$ are four vectors in space, and $IJK \neq 0$, one can determine numbers* xyz *such that one obtains the result*

$$U = xI + yJ + zK.$$

In fact, taking at random the point O, one sets $A - O = I, B - O = J$, $C - O = K, P - O = U$; from P produce the parallel to OC that intersects the plane OAB in Q. One has $P - O = (P - Q) + (Q - O)$. Now $Q - O$ lies in the plane OAB, and thus one can put

$$Q - O = x(A - O) + y(B - O);$$

$P - Q$ is parallel to $C - O$, so one can put $P - O = z(C - O)$; upon substituting one has $P - O = x(A - O) + y(B - O) + z(C - O)$, which is the formula to be demonstrated.

Theorem 2. *If $IJKU$ are any vectors whatever, one always has $IJKU = 0$.*

Theorem 3. *If ABCDP are any five points in space, and $ABCD \neq 0$, one can determine numbers* xyzt *for which there results* $P = xA + yB + zC + tD$.

Theorem 4. *For five points ABCDP in space the relation*

$$ABCD - ABCP + ABDP - ACDP + BCDP = 0$$

holds, which one can also write

$$ABCD = (ABC - ABD + ACD - BCD)P.$$

The demonstrations of these theorems are analogous to those of their correspondents in the preceding number.

Applications

19. Substituting for the expression "formation of the first species" its meaning that results from the definition given, and for greater clarity restricting oneself to all formations in which all the points have the unit for coefficient, the propositions of N. 13 can be expressed in the following way:

1. If $A_1, A_2, \dots, A_n$ are n points in space, one can determine a point G, their barycenter, which has the property that, however one takes the triangle PQR, one always has

$$A_1PQR + A_2PQR + \dots + A_nPQR = nGPQR,$$

the volumes being considered in magnitude and sense.

2. Let U be a point in space, whose distance from the plane PQR is the unit of measure; then the ratio $\frac{APQR}{UPQR}$ represents the distance of the point A from the plane PQR, taking the sign $+$ if A and U stand on the same side of the plane, with the sign $-$ if they are opposite. Dividing all the terms of the preceding identity by $UPQR$, one deduces:

The sum of the distances of the points $A_1 \dots A_n$ from a plane is equal to n times the distance of the barycenter G of those points from that plane, which distance is taken positively for all the points that stand on one side, chosen by convention, of the plane, negatively for the points that stand on the opposite side. This sum of the distances is zero for all the planes that pass through G, and is constant, in absolute value, for the planes tangent to a fixed sphere with center G.

3. Let $A_1 \dots A_m$, $B_1 \dots B_n$; $C_1 \dots C_p$, $D_1 \dots D_q$ be four systems of points. Given G, H, K, L as their barycenters, whence $A_1 + \dots A_m = mG$,

$B_1 + \ldots + B_n = nH$, $C_1 + \ldots + C_p = pK$, $D_1 + \ldots + D_q = qL$, one has

$$[(A_1 + \ldots A_m)(B_1 + \ldots + B_n)(C_1 + \ldots + C_p)(D_1 + \ldots + D_q) = mnpqGHKL,$$

or: The sum of all the tetrahedrons having for first vertex any point of the first system, for second vertex any point of the second system, and so on, is equal to the tetrahedron having for vertices the barycenters of the four given systems, multiplied by mnpq.

4. Again setting $nG = A_1 + A_2 + \ldots + A_n$, if one divides the system of given points into two groups, one of which contains r points and the other the remaining $n - r$, given G_1 and G_2 as barycenters of these groups, one has $nG = rG_1 + (n - r)G_2$; whence the point G is on the line $G_1 G_2$, and divides it into parts inversely proportional to the numbers r and $n - r$. Varying the groups into which the system can be decomposed, one has $2^{n-1} - 1$ right-lines passing through the barycenter.

Setting $n = 3$ and $n = 4$ one has the propositions:
The medians of a triangle pass through a common point.
The right-lines that in a tetrahedron join the vertices with the barycenters of the opposite faces, and the right-lines that join the median points of the opposite corners pass through a common point.

5. To construct the triangle ABC knowing the barycenter G, the point H that divides the side AC into parts that stand in the ratio $2 : 1$, and the point K on the prolongation of BC at a distance from C equal to BC.
The preceding conditions can be transformed into the equations:

$$A + B + C = 3G$$
$$H - A = 2(C - H)$$
$$K - C = C - B.$$

These three equations, treated as linear algebraic equations in A B C determine those unknowns. Eliminating A and B one recovers

$$C = K + 3(G - H)$$

whence

$$B = C + (C - K)$$

and finally

$$A = H + 2(H - C).$$

whereby one has the simplest construction of the three points requested.

6. If I is a fixed vector, the equivalence

$$P' = P + I$$

makes correspond to every point P in space a point P' called the correspondent of P under the *translation* determined by the vector I.

7. If O is a fixed point, and k a given number, the equivalence

$$P' = O + k(P - O)$$

makes correspond to every point P a point P' which is called *homothetic* to P with center of homothety in O and with respect to the coefficient of homothety k.

8. If P_1 is the correspondent of P under the translation I, that is $P = P_1 + I$, and if P_2 is the correspondent of P_1 under the homothety (O, k), that is $P_2 = O + k(P_1 - O)$, then

(1) $$P_2 = O + kI + k(P - O)$$

is the relation between P and P_2. One determines the point C according to

(2) $$C = O + kI + k(C - O);$$

one has

$$C = O + \tfrac{k}{1-k}I.$$

Subtracting (2) from (1) one deduces

$$P_2 = C + k(P - C),$$

that is, P_2 is homothetic to P with center of homothety in C and with respect to k.

9. If P_1 is homothetic to P with center O and with respect to k, that is

$$P_1 = O + k(P - O),$$

and if P_2 is homothetic to P_1, with center O_1 and with respect to k_1, that is

$$P_2 = O_1 + k_1(P_1 - O_1),$$

one has between P and P_2 the relation

(1) $$P_2 = O_1 + k_1(O - O_1) + kk_1(P - O).$$

One determines the point C for which

(2) $$C = O_1 + k_1(O - O_1) + kk_1(C - O)$$

from which one recovers

$$C = O + \tfrac{1-k_1}{1-kk_1}(O_1 - O)$$

and C is effectively determined if $1 - kk_1$ is not zero.

Subtracting (2) from (1) one deduces

$$P_2 = C + kk_1(P - C),$$

whence P_2 is homothetic to P with center in C and with respect to kk_1.

What happens if $kk_1 = 1$?

10. If $A\,B\,C$ are three points on a straight line, and A and B do not coincide, one has

$$C = \frac{CB}{AB}A + \frac{AC}{AB}B.$$

11. Let $A\,B\,C$ be three vertices of a triangle. Let one take upon its sides the points $A_1 B_1 C_1$. One has

$$A_1 = \frac{A_1 C}{BC}B + \frac{BA_1}{BC}C, \quad B = \frac{B_1 A}{CA}C + \frac{CB_1}{CA}A, \quad C = \frac{C_1 B}{AB}A + \frac{AC_1}{AB}B$$

whence, carrying out the calculation,

$$A_1 B_1 C_1 = \left(\frac{A_1 C}{BC} \cdot \frac{B_1 A}{CA} \cdot \frac{C_1 B}{AB} + \frac{BA_1}{BC} \cdot \frac{CB_1}{CA} \cdot \frac{AC_1}{AB} \right) ABC.$$

Supposing ABC is not zero, the necessary and sufficient condition for the three points $A_1 B_1 C_1$ to be on a straight line, whence $A_1 B_1 C_1 = 0$, is that the coefficient of ABC be set to zero. Thus if the points $A_1 B_1 C_1$, which lie on the sides BC, CA, AB of the triangle ABC, are in a straight line, one has

$$\frac{BA_1}{A_1 C} \cdot \frac{CB_1}{B_1 A} \cdot \frac{AC_1}{C_1 B} = -1$$

and vice-versa. This is the *theorem of Menelaus*.

12. Let A, B, C, D be the successive vertices of an oblique {i.e., nonplanar} quadrilateral. Let one take upon its sides AB, BC, CD, DA the points P, Q, R, S. One can calculate volume of the tetrahedron $PQRS$ in functions of $ABCD$, and of the ratios of the segments into which the points $PQRS$ divide the sides, in a way analogous to the preceding, and one finds

$$PQRS = \left(\frac{PB}{AB} \cdot \frac{QC}{BC} \cdot \frac{RD}{CD} \cdot \frac{SA}{DA} + \frac{AP}{AB} \cdot \frac{BQ}{BC} \cdot \frac{CR}{CD} \cdot \frac{DS}{DA} \right) ABCD.$$

Thus, supposing $ABCD$ is not zero, the necessary and sufficient condition for the points $PQRS$ to lie in a plane is

$$\frac{AP}{PB} \cdot \frac{BQ}{QC} \cdot \frac{CR}{RD} \cdot \frac{DS}{SA} = -1.$$

CHAPTER III

Formations
of the Second Species

20.   Definition. We call the *vector of a line AB* the vector $B - A$.

Definition. We call the *magnitude of a vector $B - A$* the magnitude of the line AB.

Problem. To construct the product AI of the point A with the vector I.

One determines $B = A + I$, whence $AB = AI$.
Thus:
Every line can be considered as the product of its origin with its vector.

Problem. Given the line AB and the point C on the right-line AB, to construct a line having C for origin and equal to AB.

If one sets $D - C = B - A$, one recovers $CD = C(B - A)$; now, since ABC are on a straight line, one has $C(B - A) = AB$, that is to say $CD = AB$.
One deduces from this that:
Any line can be transformed into another having for origin an arbitrary point of the right-line that contains the given line, and for vector that of the given line.

Problem. To construct a line equal to mAB, where A and B are points, and m is a number.

One sets $C = A + m(B - A)$, thus $AC = mAB$.

Corollary. *Every formation of second species of the form* $ma + nb + \dots$ *is reducible to a sum $a' + b' + \dots$ of several lines having coefficients equal to unity.*

Problem. To construct the sum of several lines $OA + OB + OC$ having a common origin.

One constructs the point X for which

$$X - O = (A - O) + (B - O) + (C - O)$$

whence

$$OX = OA + OB + OC.$$

Consequently

The sum of several lines having a common origin is a line having the same origin, and whose vector is the sum of the vectors of the given lines.

21. *Problem.* To construct the sum of two lines AB and $A'B'$ whose right-lines intersect at a point O.

One constructs the lines $OP = AB$ and $OP' = A'B'$; then $OQ = OP + OP'$; thus $OQ = AB + A'B'$.

Consequently

The sum of two lines that intersect at a point is represented by a line passing through the point of intersection of the given lines, and whose vector is the sum of their vectors.

Problem. To construct the sum of two parallel lines, the sum of whose vectors is not zero.

Let the lines be AI and BJ, having for origins A and B, and for vectors I and J, parallel. One takes at random, in the plane of the two lines, a line CK, whose vector K is not parallel to I and J. One has

$$AI + BJ = (AI + CK) + (BJ - CK).$$

The lines AI and CK intersect; therefore their sum is reducible to a single line having for vector $I + K$; if D is the origin of this right-line, one has $AI + CK = D(I + K)$. Analogously the sum $BJ - CK$ is reducible to a single line; if E is its origin, one has $BJ - CK = E(J - K)$, and

$$AI + BJ = D(I + K) + E(J - K).$$

The two lines $D(I + K)$ and $E(J - K)$ are not parallel. Indeed, if they were parallel, one would have $(I + K)(J - K) = 0$, that is

$$IJ - IK - JK = 0;$$

and since $IJ = 0$, one would have $(I + J)K = 0$, which cannot be, since $I + J$ is not zero by hypothesis, and its direction, which is that of I and J, is different from that of K. Consequently these two lines can be reduced to a single line having for its vector $I + K + J - K = I + J$.

The auxiliary line CK can be taken at random; if one takes it to be AB, one has the following construction; One sets $B' - B = I$, $A' - A = J$; one draws the lines $AB' = AI + AB$ and $BA' = BJ + BA$.

Thus $AI + BJ = AB' + BA'$; in other words the line sought passes through the point O of intersection of AB' and BA'; knowing the vector $I + J$, it is immediately constructed.

Consequently

The sum of two parallel lines, where the sum of their vectors is not zero, is reducible to a single line whose vector is the sum of the given vectors; it is therefore parallel to those lines.

Problem. To construct the sum of several lines lying in the same plane, supposing the sum of their vectors is not zero.

The construction has already been performed for the case of two lines. If the number of lines is > 2, one considers three of them at random; their vectors cannot simultaneously be equal and of opposite sign pairwise; thus, of those three, one takes whichever two whose vectors are not equal and of opposite sign, and substitutes for their sum the line to which it is equal, and which one has constructed. In this way one has reduced the proposed sum to another containing one line less; and the sum of the vectors of the second formation is equal to that of the first. Continuing thus one reduces the system to two lines, and then to one. Thus

The sum of several lines lying in the same plane, where the sum of the vectors of those lines is not zero, is reducible to a single line whose vector is the sum of the vectors of the given line.

Corollary. *The sum of several lines lying in the same plane, where the sum of their vectors is zero, is reducible to the sum of two lines having their vectors equal and of opposite sign. One of the two lines one can take at random in the plane.*

In fact, let the sum be $a_1 + a_2 + \ldots + a_n$, and let the sum of the vectors of these lines be zero. Taking at random in the plane the nonzero line p, the sum $a_1 + a_2 + \ldots + a_n + p$ is such that the sum of the vectors of the lines of which it is composed is equal to the vector of p, which is not zero; thus that sum is reducible to one line

$$q = a_1 + a_2 + \ldots + a_n + p;$$

and the vector of q is equal to that of p. One recovers from this

$$a_1 + a_2 + \ldots + a_n = q - p,$$

and thus the proposed sum is reduced to the sum of the two lines q and $-p$, having vectors equal and of opposite signs.

22.   The construction of the sum line, or resultant, of several lines lying in the same plane can be arranged practically in this way. For convenience in writing let

a and b be two lines in the plane that intersect in a point, with ab understood as their point of intersection (*vide* N. 41).

Let $a_1 a_2 a_3 a_4$ be the given lines, $l_1 l_2 l_3 l_4$ their vectors. Taking the point O in the plane at random, one constructs the points $A_1 A_2 A_3 A_4$ that satisfy all the following conditions:

$$A_1 - O = l_1, \quad A_2 - A_1 = l_2, \quad A_3 - A_2 = l_2, \quad A_4 - A_3 = l_4.$$

One has

$$A_2 - O = l_1 + l_2, \quad A_3 - O = l_1 + l_2 + l_3, \quad A_4 - O = l_1 + l_2 + l_3 + l_4.$$

The line $a_1 + a_2$ passes through the point $a_1 a_2$, and has for its vector $l_1 + l_2 = A_2 - O$; then from the point $a_1 a_2$ one produces the parallel to OA_2; upon this falls the line $a_1 + a_2$.

The line $a_1 + a_2$ intersects a_3 in the point $(a_1 + a_2)a_3$. From this point one produces the parallel to OA_3; upon this right-line falls the line $a_1 + a_2 + a_3$. This right-line intersects a_4 in the point $(a_1 + a_2 + a_3)a_4$; if from this point one produces the parallel to OA_4, upon this last right-line falls $a_1 + a_2 + a_3 + a_4$; and since the vector of this line equals $A_4 - O$, the line sought is completely determined.

In this construction it is implicitly supposed that the right-lines a_1 and a_2, $a_1 + a_2$ and a_3, $a_1 + a_2 + a_3$ and a_4 are not parallel, that is, no line of the polygon $OA_1 A_2 A_3 A_4$, save the first, passes through O. If this condition is not satisfied (or, in practice, even though it is satisfied, if the right-line on which one is to determine the points of intersection is intersected in too small an angle, or beyond the page), one can construct the resultant in the following way.

Construct, as before, the polygon $OA_1 A_2 A_3 A_4$, taking at random in the plane the point P, provided it does not fall on any side of this polygon; let p be a line having for its vector $O - P$, and construct according to the preceding rule the resultant of the system $p + a_1 + a_2 + a_3 + a_4 - p$, which is equal to that given.

Thus from the point pa_1 one produces the parallel to PA_1 (which contains $p + a_1$); this intersects a_2 in the point $(p + a_1)a_2$; from this point one produces the parallel to PA_2 (which contains $p + a_1 + a_2$): this intersects a_3 in $(p + a_1 + a_2)a_3$; from this point one produces the parallel to PA_3 (which contains $p + a_1 + a_2 + a_3$); this intersects a_4 in $(p + a_1 + a_2 + a_3)a_4$; from this point one constructs the parallel to PA_4 (which contains $p + a_1 + a_2 + a_3 + a_4$); this right-line intersects $-p$ in a certain point $(p + a_1 + a_2 + a_3 + a_4)p$, through which passes the line $p + a_1 + a_2 + a_3 + a_4 - p$. Knowing then that the vector of this line equals $l_1 + l_2 + l_3 + l_4 = A_1 - O$, it is immediately determined.

The polygon having for sides $p, p + a_1, p + a_1 + a_2, p + a_1 + a_2 + a_3, \ldots$, and its vertices, as a consequence, on the given lines $a_1 a_2 a_3 \ldots$ is called, in Graphic

Statics, a *funicular polygon* that connects the given lines (or forces). A system of lines has an infinity of funicular polygons because of the arbitrariness of the line p.

23. *Problem.* To construct the sum of several parallel lines.

Let the parallel lines be $AA_1, BB_1, CC_1, \ldots$. If their number is greater than 2, it is always possible to take two of them whose vectors are not equal and of opposite signs; for their sum one can substitute a single line, parallel to that given, and thus one reduces the proposed system to another having one line less. Continuing in this way one reduces the system to two lines, and if the sum of the lines proposed is not zero, one can reduce the formation proposed to a single line. If however the sum of the vectors of the given line is zero, calling XX_1 a line parallel to those given, one has $AA_1 + BB_1 + \ldots = (AA_1 + BB_1 + \ldots + XX_1) - XX_1$. The sum within the parentheses is reducible to a single line having for its vector that of XX_1; thus the formation proposed is reduced to a sum of two lines with vectors equal and of opposite signs.

One arrives at the same result via this alternative route:

Let I be a vector parallel to all the lines of the system. One can determine the numbers a, b, $\ldots$ for which $A_1 - A = aI, B_1 - B = bI, \ldots$. One deduces $AA_1 - aAI, BB_1 = bBI, \ldots$; and summing,

$$AA_1 + BB_1 + \ldots = (aA + bB + \ldots)I.$$

Now if the sum $a + b + \ldots$ is not zero, that is to say, multiplying it by the vector I, if the sum of the vectors $(A_1 - A) + (B_1 - B) + \ldots$ is not zero, one determines the point G that satisfies the condition

$$(a + b + \ldots)G = aA + bB + \ldots .$$

Thus one has $AA_1 + BB_1 \ldots = (a + b + \ldots)GI = G(aI + bI + \ldots) = G[(A1 - A) + (B1 - B) + \ldots]$.

If however $a + b + \ldots = 0$, the sum $aA + bB + \ldots$ is reducible to a vector J, and one has $AA_1 + BB_1 + \ldots = JI$. Thus

The sum of several parallel lines, if the sum of the vectors of those lines is not zero, is reducible to a single line parallel to those given, having for vector the sum of their vectors, and for its origin one can take the barycenter of the origins of the given lines to which are affixed numbers proportional to their vectors. If however the sum of the vectors is zero, the formation proposed is reducible to the sum of two lines with vectors equal and of opposite signs.

24. In the theorems of N. 21 and 23 there already appeared the sum of two lines having their vectors equal and of opposite sign. Calling A and B the origins of the two lines, J the vector of the first, and thus $-J$ that of the second, the sum being considered can be written $AJ - BJ$, or $(A - B)J$; in other words it is reducible to the product of the two vectors $A - B$ and J. Vice-versa, the product IJ of two vectors I and J; where one sets $I = A - B$, becomes equal to $(A - B)J = AJ - BJ$; in other words one transforms it into the sum of two lines having vectors equal and of opposite signs. The product of two vectors, or the sum of two lines with vectors equal and of opposite sign, has, in the study of the formations of the second species, the same importance as that of vectors in formations of the first; for this reason it is useful to give it a name.

Definition. The product of two vectors is called a *bivector*.

Now from the propositions of N. 21 and 23 one deduces:

The sum of several lines lying in the same plane, and the sum of several parallel lines, when the sum of the vectors of those lines is zero, is reducible to a bivector.

A bivector IJ, that is, the product of the two vectors I and J, can be represented, as one has seen, as the sum of the two lines having their vectors equal and of opposite signs. It can also be represented in another way. Taking at random the point A, and setting $B - A = I, C - A = J$, one has $IJ = (B - A)(C - A)$, or, developing:

$$IJ = AB + BC + CA,$$

and thus the bivector is reduced to the sum of the three sides of a triangle.

The necessary and sufficient condition for a bivector to be zero is that the two vectors of which it is the product be parallel.

Definition. We say that the vector I is parallel to the bivector JK if I, J, K are parallel to the same plane, or if $IJK = 0$. We say that two bivectors IJ and $I'J'$ are parallel or that they have the same position {giacitura}, if the four vectors I, J, I', J' are parallel to the same plane.

25. *Problem.* Given the bivector IJ, and the vector K parallel to IJ, and nonzero, to determine a vector L such that $IJ = KL$.

Taking at random the point O, one sets $A - O = I, B - O = J, C - O = K$; the points O, A, B, C lie in the same plane, and one has $IJ = (A - O)(B - O) = OA + AB + BO$. One constructs the line $OA + AB + BO - OC$, which is possible since those lines lie in the same plane, and the sum of their vectors is $-(C - O)$, which is not zero; calling D any point of this line, one can cast it into the form

$$OA + AB + BO - OC = -D(C - O),$$

whence one recovers

$$OA + AB + BO = OC - D(C - O) = O(C - O) - D(C - O) = (C - O)(D - O);$$

thus, setting $L = D - O$, one has $IJ = KL$.

Problem. To construct the sum of the two bivectors i and j, given anywhere in space.

One takes a vector U parallel to the two bivectors given, and determines the vectors I and J for which $UI = i$, $UJ = j$; setting $K = I + J$, one has $UK = UI + UJ = i + j$.

Corollary. *The sum of as many bivectors in space as one wishes is reducible to a single bivector.*

26.   Theorem. *The sum of several lines in space is reducible to one line, having for origin an arbitrary point and for vector the sum of the vectors of the given line, plus a bivector.*

Let the sum $s = A_1 I_1 + A_2 I_2 + \ldots + A_n I_n$, in which $A_1 A_2 \ldots$ are the origins, $I_1, I_2, \ldots$ the vectors of the lines to be summed. Taking at random the point O, one has the identity

$$s = O(I_1 + I_2 + \ldots + I_n) + [(A_1 - O)I_1 + (A_2 - O)I_2 + \ldots + (A_n - O)I_n].$$

The first term of s represents a line having for origin the arbitrary point O, and for vector the sum of the vectors of the given lines; the second term, which is the sum of n bivectors, is reducible to a single bivector; thus the theorem is proven.

Corollary. *The sum of several lines in space, where the sum of the vectors of those lines is zero, is reducible to a bivector (which may also be set to zero).*

Thus in the preceding formula the first term disappears.

Theorem. *If the sum of the vectors of the given lines is not zero, but the sum of the volumes ss is zero, the sum of the given lines is reducible to a single line, and vice-versa.*

In fact, if one writes the formation $s = a + i$, where a represents the line, i the bivector to whose sum s can be reduced, one has

$$ss = (a + i)(a + i) = 2ai.$$

Now if $ss = 0$, one must have $ai = 0$, whence the line a is parallel to the bivector i. One therefore determines the line b such that $b - a = i$; one has $b = a + i = s$.

Vice-versa, if s is reducible to a single line b, one has $ss = bb = 0$.

Theorem. *Every sum s of lines is reducible in infinitely many ways to the sum of two lines.*

In fact, taking at random the line p, one considers the sum $s' = s - kp$, where k is a number to be determined. One has $s's' = ss - 2ksp$. Now if sp is not zero, upon setting $k = \frac{ss}{2sp}$, one has $s's' = 0$. As a result, if the sum of the vectors of s' is not zero (which is certain if the direction of the vector of p noes not coincide with the direction of the sum of the vectors of s), the sum s' is reducible to a single line q; thus $s - kp = q$, or $s = kp + q$; thus the proposed sum is reduced to the sum of two lines.

Applications

27. **1.** If A_1B_1, A_2B_2, A_3B_3, ... are lines lying in the same plane, the sum of whose vectors is not zero, and XY is the line equal to their sum, that is

$$XY = A_1B_1 + A_2B_2 + A_3B_3 + \ldots ,$$

taking at random a point P in the plane, one has

$$PXY = PA_1B_1 + PA_2B_2 + PA_3B_3 + \ldots ,$$

that is

Given several lines A_1B_1, A_2B_2, ... in a plane, if the sum of their vectors is not zero, one can construct a line XY for which, however one may take the point P in the plane, one always has

$$PXY = PA_1B_1 + PA_2B_2 + \ldots .$$

The locus of the points P for which the sum of those triangular areas, having their vertices in P and A_1B_1, A_2B_2, ... for bases, is zero is the right-line XY; the locus of the points P for which that sum has a constant value (in magnitude and sign) is a right-line parallel to XY.

If however the sum of the vectors of the lines A_1B_1, A_2B_2, ... is zero, that sum is reducible to a bivector $XY + YZ + ZX = A_1B_1 + A_2B_2 + \ldots$; multiplying by P one has

$$XYZ = PA_1B_1 + PA_2B_2 + \ldots ,$$

or in other words

If the sum of the vectors of the lines A_1B_1, A_2B_2, ... is zero, however one may take the point P in the plane, the sum of the triangles $PA_1B_1 + PA_2B_2 + \ldots$ has a constant value.

2. The lines considered can be represented as the successive sides $A_1A_2 + A_2A_3 + \ldots + A_{n-1}A_n$ of a polygonal line, open or closed. The sum of the vectors of thes lines, that is $(A_2 - A_1) + (A_3 - A_2) + \ldots + (A_n - A_{n-1}) = A_n - A_1$, is not zero if the line is open, and is zero if the line is closed. Thus in these cases the preceding proposition becomes

Given an open polygonal line $A_1A_2 \ldots A_n$, one can construct a line XY whose vector equals $A_n - A_1$, such that, however one takes the point P in the plane, one has

$$PXY = PA_1A_2 + PA_2A_3 + \ldots PA_{n-1}A_n.$$

Given a closed polygonal line $A_1A_2 \ldots A_{n-1}A_n$, however one may take the point P in the plane, the sum $PA_1A_2 + PA_2A_3 + \ldots PA_{n-1}A_n$ has a constant value.

To this constant sum one might give the name of the *enclosed area of the polygonal line $A_1A_2 \ldots A_{n-1}A_1$.* These propositions also hold if for the polygonal line one substitutes a curve.

3. The construction just made permits one to solve the following problem:

Given a closed polygonal line $A_1A_2 \ldots A_{n-1}A_1$, and in its plane a line PQ, to construct the triangle PQR equal to the enclosed area of that polygonal line.

One constructs the line $RS = A_1A_2 + A_2A_3 + \ldots + A_{n-1}A_1 - PQ$ whose vector is equal and of sign opposite to that of PQ. Then

$$A_1A_2 + A_2A_3 + \ldots + A_{n-1}A_1 = PQ + RS.$$

One multiplies by any point of RS, e.g., R, and one has

$$(A_1A_2 + A_2A_3 + \ldots + A_{n-1}A_1)R = PQR$$

whence PQR is the triangle sought.

4. If a and b are two lines passing through the same point, and equal in magnitude, the lines $a + b$ and $a - b$ represent the internal and external bisectors of the angle formed by the two given lines. If the lines a and b are parallel, equal in magnitude, and turn in the same sense, $a + b$ represents the line equidistant from the two given, $a - b$ a bivector.

5. Let ABC be a triangle; let abc be three lines lying on the sides BC, CA, AB, having the senses of those sides, and in magnitude mutually equal. Then the lines $a - b, b - c, c - a$ are on the internal bisectors of the triangle, and $a + b, b + c$, $c + a$ are on the externals. The identities

$$a + b + c = (a + b) + c = (b + c) + a = (c + a) + b$$

say that *the points of intersection of the external bisectors of a triangle with the opposite sides lie on the same right-line.*

One also has the identities

$$a + b - c = (a + b) - c = (a - c) + b = (b - c) + a$$
$$a - b + c = (a - b) + c = (a + c) - b = (c - b) + a$$

$$-a + b + c = (b - a) + c = (c - a) + b = (b + c) - a$$

which say that *the points of intersection of the two internal bisectors and of the remaining external one with the opposite side, lie on the stright line.*

The identity $a - c = (a - b) + (b - c)$ says that the internal bisector $a - c$ passes through the point of intersection of the other two, $a - b$ and $b - c$, whence *the internal bisectors of the triangle intersect in a point.*

The identities

$$a - b = (a + c) - (b + c)$$
$$a - c = (a + b) - (b + c)$$
$$b - c = (a + b) - (a + c)$$

say that *through the point of intersection of the two external bisectors also passes the other internal bisector.*

6. Let a, b, c, d be four lines lying in the same plane, and equal in magnitude; they form a *complete quadrilateral*. The identities

$$a + b + c + d = (a + b) + (c + d) = (a + c) + (b + d)$$
$$= (a + d) + (b + c) = (a + b + c) + d = \ldots$$

say that the right-line $a + b + c + d$ contains the three points of intersection of the internal bisectors of the opposite angles of the quadrilateral, and the four points of intersection of one side of the quadrilateral, e.g., d, with the right-line $a + b + c$, whose meaning in the trilateral a, b, c was already found in the preceding.

Upon changing the signs of any of the lines a, b, c, d, one has in all the 8 lines $a \pm b \pm c \pm d$, which have properties analogous to those of the line $a + b + c + d$.

7. Let $a_1 a_2 \ldots a_n$ be lines lying in the same plane, and in magnitude equal to the unit of measure. Let ω be a triangle lying in the same plane, having base and altitude equal to the unit of measure. Then, given any point P of the plane, the ratio $\frac{Pa}{\omega}$ represents the altitude of the triangle Pa, that is the distance of the point P from the right-line a, this distance being taken positive if Pa has the same sense as ω, negative in the opposite case.

If the sum $a_1 + a_2 + \ldots + a_n$ is reducible to a single line that we put into the form ma, where m is a number and a a line equal in magnitude to the unit of measure, one has

$$\frac{Pa_1}{\omega} + \frac{Pa_2}{\omega} + \ldots + \frac{Pa_n}{\omega} = m\frac{Pa}{\omega},$$

that is, the sum of the distances of the point P from the given line, those distances being taken with the conventional signs, is equal to the distance of the same point from the right-line $a_1 + a_2 + \ldots + a_n$, multiplied by the number m. This sum is zero for the points of this right-line, and has a constant value for the points that lie on a parallel to it.

If however the sum $a_1 + a_2 + \ldots + a_n$ is reducible to a bivector, the sum of the distances of the point P from the given line has a value independent of the position of P.

In the plane, the locus of points for which the sum of the distances from the given right-lines $a_1 \ldots a_n$ is a constant, those distance being taken in absolute value, form the perimeter of a polygon whose vertices lie on the given right-lines, and whose sides are parallel to the right-lines $\pm a_1 \pm a_2 \pm \ldots \pm a_n$.

8. Let a, b, c be three lines equal in magnitude passing through the same point O, but not lying in the same plane. One can consider the right-lines a, b, c as the edges of a trihedral angle. The identities

$$a + b + c = (a + b) + c = (a + c) + b = (b + c) + a$$

say that the three planes that pass through an edge, e.g., of the trihedron, and through the internal bisector $b + c$ of the opposite face, all contain the right-line $a + b + c$.

Analogously, through each of the right-lines $a + b - c, a - b + c, -a + b + c$ pass three planes that connect an edge of the trihedron with a bisector of the opposite face.

The identities $(a - b) + (b - c) + (c - a) = 0$, $(a + b) - (b + c) - (c - a) = 0$, $(b + c) - (c + a) + (a - b) = 0$, $(c + a) - (a + b) + (b - c) = 0$ say that the three external bisectors lie in the same plane, as also two internal bisectors and the other external one.

9. Given a polygonal line $A_1 A_2 \ldots A_n$ in space, one can always determine, in infinitely many ways, two lines PQ and $P'Q'$ for which, however one takes the points XY, the volume described by the triangle MXY, where the point M describes the polygon $A_1 \ldots A_n$, is equal to the sum $PQXY + P'Q'XY$.

10. We will demonstrate in the sequel that, projecting on a plane, with parallel projections, two systems of equal lines, their projections are also equal. One then deduces from the theorem demonstrated that

Given any closed polygonal line whatever (not planar), one can always determine a triangle ABC for which, projecting on an arbitrary plane with parallel projections of arbitrary direction both the given polygonal line and the triangle ABC, the enclosed area of the projection of the closed polygon is equal to the projection of the triangle ABC.

11. In Mechanics one represents a *force* with a line; a *system of forces* with a formation of the second species; two systems of forces applied to the same rigid body are *equivalent* if they are represented by equal formations of the second species. The system of forces represented by a bivector, that is by two lines having vectors equal and of opposite signs is called a *force couple*. From the propositions demonstrated there results:

Every system of forces applied to the same point is reducible to a single force; every system of forces lying in a plane is reducible to a single force or to a couple; every system of couples is reducible to a single couple; every system of forces is reducible in infinitely many ways to a force and a couple together.

If a is a line equal in magnitude to the unit of measure, and f is another line representing a force, the volume af (or better, the number that measures it) is called in Mechanics the *moment of the force f with respect to the axis a*. If two systems of forces s and s' are equivalent, that is $s = s'$, one deduces $as = as'$, that is *the sum of the moments of the forces of two equivalent systems with respect to an arbitrary axis are equal*.

From the last theorem of N. 23 there results:

Every system of forces is reducible in infinitely many ways to the sum of two forces; one can take at random the right-line that contains one of the two forces, provided the sum of the moments of the forces of the system with respect to that right-line is not zero, and provided the right-line is not parallel to the sum of the vectors of the given forces.

CHAPTER IV

Formations
of the Third Species

28.   Definition. We call the *bivector* of a triangular surface ABC the bivector $AB + BC + CA$, that is $(B - A)(C - A)$.

Definition. We call the *magnitude of a bivector* $AB + BC + CA$ the magnitude of ABC.

Problem. To construct the product of a point A by a bivector IJ.

Setting $B = A + I$, $C = A + J$, one has $AIJ = A(B-A)(C-A) = ABC$; thus the proposed product is a triangle, one vertex of which is A, and whose bivector is equal to the bivector given.

Problem. Given the triangle ABC, and in its plane the line PQ, to construct the point R for which $ABC = PQR$.

One constructs the line $RS = AB + BC + CA - PQ$, which lies in the plane $ABCPQ$, and one has $AB + BC + CA = PQ + RS$.

On RS taking a point at random, e.g., R, one has

$$(AB + BC + CA)R = PQR,$$

that is

$$ABC = PQR.$$

Problem. To construct the surface $mABC$, where m is a number.

Setting $B' - A = m(B - A)$, one has $AB'C = mABC$.

29.   *Problem.* To construct the sum of two surfaces $\alpha + \beta$.

If the planes of the two surfaces intersect, one takes at random on their intersection the line PQ; one constructs the points M and N for which $PQM = \alpha$, $PQN = \beta$. One has

$$\alpha + \beta = PQ(M - P) + PQ(N - P).$$

Setting $R - P = (M - P) + (N - P)$, one has

$$\alpha + \beta = PQR.$$

One observes that the bivector of PQR, that is $(Q - P)(R - P)$, is equal to $(Q - P)(M - P) + (Q - P)(N - P)$, that is it is the sum of the bivectors of the given surfaces.

If the planes of the two surfaces are parallel, one takes in the first plane two points A and B, and in the second another two A' and B' such that $B' - A' = B - A$; {in the first plane} one constructs the point C such that $ABC = \alpha$, and in the second the point C' such that $A'B'C' = \beta$, and in addition for which the right-lines AC and $A'C'$ are parallel. Suppose that $(C - A) + (C' - A') \neq 0$; one constructs $RS = AC + A'C'$, and the point $T = R + (B - A)$. One has

$$\alpha + \beta = ABC + A'B'C' = A(B - A)C + A'(B' - A')C'$$
$$= -(AC + A'C')(B - A) = -RS(T - R) = RTS.$$

In this construction one supposed it was *not* the case that

$$(C - A) + (C' - A') = 0.$$

If this condition *is* satisfied, one also has

$$(B - A)(C - A) + (B' - A')(C' - A') = 0,$$

that is the sum of the bivectors of the given surfaces is zero. In this case $\alpha + \beta$ is the sum of two surfaces having bivectors equal and of opposite signs, and one has

$$\alpha + \beta = A(B - A)(C - A) + A'(B' - A')(C' - A') = (A - A')(B - A)(C - A),$$

that is, the sum of two surfaces having equal vectors, but of opposite sign, is the product of three vectors.

Definition. We call the product of three vectors a *trivector*.

Thus from the solution of the last problem there results the following:

Theorem. *The sum of several triangular surfaces, if the sum of the bivectors of those surfaces is not zero, is reducible to a single triangular surface. If however the sum of the bivectors of those surfaces is zero, the sum proposed is reducible to a trivector.*

30. As one has already seen, a trivector is equal to the sum of two triangular surfaces having bivectors equal and of opposite signs. One can also give another notable form for a trivector.

Consider the trivector IJK. Taking at random A, and setting $B - A = I$, $C - A = J, D - A = K$, one recovers

$$IJK = (B - A)(C - A)(D - A) = BCD - ACD + ABD - ABC$$
$$= BCD + DCA + ACB + ABD,$$

and thus is reduced to the sum of the faces of a tetrahedron, in such a way that every side is traversed twice in opposite senses.

The identity

$$P(BCD - ACD + ABD - ABC) = ABCD$$

says that the product of a point by a trivector gives a volume independent of the position of the point. Thus two trivectors are equal if, when cast in the form of the sum of the faces of two tetrahedrons, those tetrahedrons are equal.

Applications

31. **1.** Let $ABC, A'B'C', A''B''C'', \ldots$ be triangles given in magnitude and position in space. If the sum of the bivectors of these triangles is not zero, constructing the triangle $PQR = ABC + A'B'C' + \ldots$, then for any point X in space, the sum of the tetrahedrons having their vertices in X and for bases the given triangles is equal to the tetrahedron having the same vertex and for base the triangle PQR. If however the sum of the bivectors of the triangles is zero, the sum of the tetrahedrons $XABC + XA'B'C' + \ldots$ has a value independent of the position of the point X.

2. We say that a system of triangles in space, given in position and sense, forms a closed polyhedral surface if every line that is the side of a triangle, taken in the opposite sense, is also the side of a triangle. The sum of the bivectors of these triangles is zero, whence the sum of the tetrahedrons, having for vertices a point X and for bases the faces of the polyhedron, has a value independent of X. To that constant sum one can give the name the *enclosed volume of that polyhedral surface*.

3. A system of triangles in space, given in position and sense, such that the sides of those triangles, some excepted, are traversed in two opposite senses, and such that those traversed only once can be ordered so that they are the consecutive sides of a closed polygonal line l, are said to form an open polyhedral surface, having for contour the polygon l.

Supposing the bivector l, that is the bivector of the system of triangles given, is not zero, construct the triangle PQR, the sum of the faces of the polyhedron; then, whatever the point X, the volume of the polyhedral solid, with vertex X and base the given polyhedral surface, is equal to the volume $XPQR$. This volume will be zero if the point X lies in the plane PQR; it has a constant value if the point X is on a plane parallel to PQR.

4. If α and β are two triangular surfaces equal in magnitude, and intersecting, $\alpha + \beta$ and $\alpha - \beta$ represent the two plane bisectors of the dihedron $\alpha\beta$, the first internal, the second external.

Consider the trihedron with vertex O and edges OA, OB, OC; if on the planes of the faces one takes three surfaces α, β, γ equal in magnitude, and having the senses of BOC, COA, AOB, then $\alpha + \beta$, $\beta + \gamma$, $\gamma + \alpha$, and $\alpha - \beta$, $\beta - \gamma$, $\gamma - \alpha$ are the plane bisectors of the angles of the trihedron, the first three the external bisectors, the others internal. The identities

$$\alpha + \beta + \gamma = (\alpha + \beta) + \gamma = (\beta + \gamma) + \alpha = (\gamma + \alpha) + \beta$$

say that *the three right-lines of intersection of the external plane bisectors with the opposite faces all lie in the plane* $\alpha + \beta + \gamma$.

Analogously, the planes $\alpha + \beta - \gamma$, $\alpha - \beta + \gamma$, $-\alpha + \beta + \gamma$ each contain three intersections of the plane bisectors of the trihedron with the opposite faces.

The identity

$$(\alpha - \beta) + (\beta - \gamma) + (\gamma - \alpha) = 0$$

says that *the three internal plane bisectors pass through the same right-line*. The identities recovered from this one upon changing the sign of one of the surfaces α, β, γ, say that two external plane bisectors and the other internal one also pass through the same right-line.

5. Let $ABCD$ be a tetrahedron, and let α, β, γ, δ be four surfaces, equal in magnitude, lying on the faces of the tetrahedron, and having the senses of BCD, $-ACD$, ABD, $-ABC$.

The tetrahedron has six dihedra and 12 plane bisectors that one can represent with $\alpha \pm \beta$, $\alpha \pm \gamma$, $\alpha \pm \delta$, $\beta \pm \gamma$, $\beta \pm \delta$, $\gamma \pm \delta$. Every one of these planes intersects the opposite edge at a point, and thus one has 12 points P. One demonstrates with an argument analogous to the preceding that there exist 16 right-lines p, every one of which contains three points P, and 8 planes π, each of which contains 4 right-lines p and six points P. The 4 right-lines p lying in a plane π are the sides of a complete quadrilateral for which the 6 points P are the vertices.

CHAPTER V

Formations on a Right-Line

32. In this chapter we will consider only points that lie on a fixed right-line, and the formations that one can make with these points; we need consider only formations of the first and second species, since those of higher species are all zero.

We will fix at random a line u, which we will give the name of the *unit line*; we will indicate by U the vector of the line u, and call it the *unit vector*. Thus, if O is an arbitrary point, between the two units U and u there is the relation $OU = u$.

If A is a formation of the first species $A = m_1 A_1 + m_2 A_2 + \ldots + m_n A_n$, in which $m_1 \ldots m_n$ are numbers, $A_1 \ldots A_n$ points, one has $AU = m_1 A_1 U + \ldots + m_n A_n U$; and since $A_1 U = A_2 U = \ldots = A_n U = u$, one has $AU = (m_1 + \ldots + m_n)u$, or again

$$\frac{AU}{u} = m_1 + m_2 + \ldots + m_n,$$

that is, $\frac{AU}{u}$ represents the sum of the coefficients of the points that constitute A, or the mass of the system A. The equation $AU = O$ signifies that the formation A is reducible to a vector. If AU is not zero, the expression $\frac{u}{AU} A$ represents the simple point to which, multiplied by the mass of A, the system A is reducible.

If a is a line, the expression $\frac{a}{u} U$ represents the vector of a.

33. To simplify the notation, when one has fixed the unit line u, instead of $\frac{a}{u}$ we will simply write a; thus aU represents the mass of the system A; aA represents the formation A multiplied with respect to $\frac{a}{u}$; aU represents the vector of a; ab represents the product of the two numbers $\frac{a}{u}$ and $\frac{b}{u}$.

Theorem. *If ABC are three formations of the first species (lying on a fixed right-line), one has the identity*

(1) $$AB.C + BC.A + CA.B = 0.$$

In fact, first let A, B, C be points that do not all coincide; and e.g., let A and B be distinct; then one can determine numbers x and y for which there results $C = xA + yB$. One multiplies this equation by B and by A; one deduces $CB = xAB$, $AC = yAB$; whence $x = \frac{CB}{AB}$, $y = \frac{AC}{AB}$; substituting, in the preceding formula, for x and y their values, making their denominators disappear, one has (1). Then, if the points ABC coincide, the three terms of (1) are zero, and this equivalence is also satisfied; thus (1) is demonstrated, if ABC are points.

Suppose ABC are any formations: $A = m_1 A_1 + \ldots m_h A_h$, $B = n_1 B_1 + \ldots + n_k B_k$, $C = p_1 C_1 + \ldots + p_l C_l$, where A_r, B_s, C_t are points, Then they will satisfy all the equivalences of the form

$$A_r B_s.C_t + B_s C_t.A_r + C_t A_r.B_s = 0.$$

One multiplies this identity by $m_r n_s p_t$, and sums all the identities that one recovers from this, attributing to r, s, t the values between 1 and h, 1 and k, 1 and l; one obtains formula (1).

One can also write formula (1) as

(2) $AB.C = AC.B - BC.A.$

Multiplying it by any formation D of the first species, one deduces

(3) $AB.CD = AC.BD - BC.AD$

which one can also write

(4) $AB.CD = \begin{vmatrix} AC & AD \\ BC & BD \end{vmatrix}.$

If in (1) one supposes ABC are (simple) points, and one multiplies it by the vector U (point at infinity), one deduces

(5) $AB + BC + CA = 0.$

If in (2) one sets $C = U$, one deduces

$$AB.U = AU.B - BU.A,$$

which gives the vector of the line AB expressed in functions of the two factors A and B, and of their masses.

34.   Let A_1 and A_2 be two formations of the first species, whose product is not zero; then if A is a new formation of the first species, one can always determine

(1) $A = x_1 A_1 + x_2 A_2.$

Indeed, the identity $A_1 A_2.A = AA_2.A_1 + A_1 A.A_2$ is transformed into the preceding where one sets

(2) $x_1 = \dfrac{AA_2}{A_1 A_2}, \qquad x_2 = \dfrac{A_1 A}{A_1 A_2}.$

Definition. The numbers x_1 and x_2 that satisfy (1) are called *the coordinates of the formation A* with respect to the reference formations A_1 and A_2.

One has the following propositions, in which A and B represent formations of the first species, and k is a number.

$$(A = x_1 A_1 + x_2 A_2) < (kA = kx_1 A_1 + kx_2 A_2),$$
$$(A = x_1 A_1 + x_2 A_2) \cap (B = y_1 A_1 + y_2 A_2)$$
$$< [A + B = (x_1 + y_1)A_1 + (x_2 + y_2)A_2],$$
$$(A = x_1 A_1 + x_2 A_2) \cap (B = y_1 A_1 + y_2 A_2) < [AB = (x_1 y_2 - x_2 y_1)A_1 A_2],$$
$$(A = x_1 A_1 + x_2 A_2) < (AU = x_1 A_1 U + x_2 A_2 U),$$

which give the coordinates of kA, $A + B$, the line AB, and the mass of A in functions of the coordinates of A and B (and of the line $A_1 A_2$ formed of the reference elements, and of the masses of the reference elements).

35.   Among the infinite systems of coordinates, corresponding to all the various choices of reference elements, those merit special mention in which one takes for reference elements a point O and the unit vector U. Then every formation A of first species is reducible to the form $A = mO + xU$; one has $AU = mu$, that is m is the mass of A. Every point is reducible to the form $O + xU$; the number x is called the *Cartesian coordinate of the point*. Every vector is reducible to the form xU; the number x is called the *coordinate of the vector*.

Applications

36.   1. Given the coordinates x_1 and x_2 of the formation A with respect to the reference formations A_1 and A_2, find the coordinates of A with respect to two other reference formations B_1 and B_2.

One has $A = x_1 A_1 + x_2 A_2$; in order that the problem be solvable, one must know the relations that hold between A_1, A_2 and B_1, B_2; we will suppose given the coordinates of A_1 and A_2 with respect to B_1 and B_2, and let $A_1 = m_{11}B_1 + m_{12}B_2$, $A_2 = m_{21}B_1 + m_{22}B_2$. Substituting these values of A_1 and A_2 into the expression for A, one recovers

$$A = (m_{11}x_1 + m_{21}x_2)B_1 + (m_{12}x_1 + m_{22}x_2)B_2,$$

which gives the coordinates of A with respect to the reference elements B_1 and B_2.

2. Let $ABCD$ be four elements of first species on the right-line; the number $\frac{AC}{BC} : \frac{AD}{BD}$ is called their *double ratio*, and is indicated by $(ABCD)$. It is a function of the four formations $ABCD$ which is not altered if one multiplies any one of those formations by a number; consequently that double ratio depends only on the positions of the four elements.

If ABC are points, and D is a vector (point at infinity), one has $AD = BD$; thus $(ABCD) = \frac{AC}{BC}$.

3. Four elements $ABCD$ are called *harmonic* if their double ratio equals -1, that is if $AC.BD + BC.AD = 0$. Given three elements ABC, the fourth harmonic D must satisfy the equation $(AC.B + BC.A)D = 0$; for this it suffices to set $D = AC.B + BC.A$; every other value of D that satisfies the equation proposed is obtained by multiplying the preceding by a number. If A and B are simple points, and C is a vector U, one has $AC = BC = 1$, whence $D = A + B$ is the fourth harmonic after the points A and B and the vector U.

CHAPTER VI

Formations in the Plane

Vectors

37. In this chapter we will occupy ourselves with systems of points in a fixed plane, and the formations that can be made from them. These formations can be of first, second, or third species, since those of the fourth species are zero.

We will fix at random a surface ω, to which we will give the name the *unit surface*. We will indicate by u the bivector of this surface, and call it the *unit bivector*. Thus, if O is any (simple) point in the plane, between the two units there exists the relation

$$Ou = uO = \omega.$$

If A is a formation of the first species $A = m_1 A_1 + \ldots + m_n A_n$, in which $A_1 \ldots A_n$ are points, one has $uA = (m_1 + \ldots + m_n)\omega$, or

$$\frac{uA}{\omega} = m_1 + \ldots + m_n.$$

that is, $\frac{uA}{\omega}$ represents the mass of the formation of the first species A. The equation $uA = 0$ says that A is a vector (point at infinity). If uA is not zero, $\frac{uA}{\omega} A$ represents the simple point coincident with A. If α is a surface, $\frac{\alpha}{\omega} u$ represents the bivector of the surface.

38. We will begin by concerning ourselves with vectors and bivectors lying in the plane. For notational simplicity, if i is any bivector, instead of $\frac{i}{u}$ we will read simply i.

It has been demonstrated that if the vectors I and J do not coincide, one can cast every other vector R (in the plane) into the form

$$(1) \qquad\qquad R = xI + yJ.$$

The numbers x and y are called the *coordinates of the vector R with respect to the reference vectors I and J*. Multiplying the preceding equation by I and by J one recovers, after having divided by IJ,

$$(2) \qquad\qquad x = \frac{KJ}{IJ}, \qquad y = \frac{IK}{IJ},$$

formulas which express the coordinates of K with respect to the bivector.

If, in (1), in the places of x and y one substitutes their values given by (2), one recovers, upon making the denominator disappear,

(3) $IJ.K = IK.J - JK.I,$

a relation between any three vectors in the plane.

If one multiplies (3) by a vector L, one has the identity between four vectors in the plane

(4) $IJ.KL = IK.JL - JK.IL,$

which one can also cast into the form

(5) $IJ.KL = \begin{vmatrix} IK & IL \\ JK & JL \end{vmatrix}.$

Carrying out the product on the left one has

(6) $(xI + yJ)(x'I + y'J) = (xy' - x'y)IJ,$

which expresses the bivector product of two vectors as a function of the coordinates of those vectors.

39. Until now the unit surface has not been definitely specified, and thus neither has the unit bivector. We will agree to take for the *unit bivector* the bivector product of two vectors I and J whose lengths are equal to the unit of length, and mutually orthogonal. (In addition, in figures, one can suppose, e.g., that a right-line describing the right angle IJ from I to J, is moved in the sense of the hand of a clock when viewed from a point in front of the page, that is, in such a way that if O is a point in the plane of the figure, and P is a point in front of the page, the volume $POIJ$ is right-handed). As a consequence the unit surface is OIJ. If one sets $A = O + I, B = O + J$, one has $OIJ = OAB$, that is the unit surface is a triangle having for base and for altitude the units of length, that is half the square with sides of unit length.

40.   Definition. If U is any vector in the plane, by $\perp U$ we will mean the vector equal in magnitude to U, whose direction is normal to that of U, and whose sense is such that the bivector $U \perp U$ is positive (that is, of the same sense as the unit bivector).

We will also say that one obtains $\perp U$ upon rotating U by a positive right angle. Instead of $\perp (xU)$, $x(\perp U)$, $(\perp U) + V$, $U(\perp V)$, we will write simply $\perp xU$, $x \perp U, \perp U + V, U \perp V$. The symbol $\perp$ can be read *perpendicular*.

The following formulas are obvious:

(1) $(U = U') = (\perp U = \perp U')$.

(2) $\perp\!\perp U = -U$.

(3) $\perp xU = x \perp U$.

(4) $\perp (U + V) = \perp U + \perp V$.

(5) $(\perp U)(\perp V) = UV$.

Formulas (3) and (4) express the distributive property of the operation $\perp$; from it one deduces

(4') $\perp (xI + yJ) = x \perp I + y \perp J$.

Substituting $\perp V$ into (5) in place of V, one deduces $(\perp U)(-V) = U \perp V$, that is, inverting the order of the factors of the first member, upon which the sign changes, one has

(6) $V \perp U = U \perp V$.

Thus the expression $U \perp V$ represents a number (that is the ratio of the bivector product of U and $\perp V$ with the unit bivector), which is a distributive function of the two factors, and is commutative.

Let U be any vector whatever, $mg\ U$ its magnitude; setting $\frac{U}{mg\ U} = K$, K is a vector equal in magnitude to the unit.

One deduces that $U = (mg\ U)K$, $\perp U = (mg\ U) \perp K$, $U \perp U = (mg\ U)^2 K \perp K$; now since the bivector $K \perp K$ is the unit bivector, which one has agreed to suppress, one has

(7) $U \perp U = (mg\ U)^2$.

The condition for the orthogonality of the two vectors U and V is $U \perp V = 0$.

41.   If U and V are two vectors in the plane, we will put, by definition,

(1) $\sin(U, V) = \dfrac{UV}{(mg\ U)(mg\ V)}$,   (2) $\cos(U, V) = \sin(U, \perp V)$.

One deduces that if U and V are vectors whose magnitudes are unity, their sine is equal to the bivector UV. One recovers immediately

(3) $UV = (mg\ U)(mg\ V)\sin(U, V)$.

(4) $\sin(U, U) = 0$, $\sin(U, \perp U) = 1$, $\sin(U, -\perp U) = -1$.

(5) $\sin(U, V) = -\sin(V, U) = -\sin(-U, V) = -\sin(U, -V)$
                $= \sin(-U, -V) = \sin(\perp U, \perp V)$.

(6) $$\cos(U, V) = \frac{U \perp V}{(mg\ U)(mg\ V)}.$$

(7) $$U \perp V = (mg\ U)(mg\ V)\cos(U, V).$$

(8) $$\cos(U, U) = 1, \quad \cos(U, -U) = -1, \quad \cos(U, \perp U) = 0.$$

(9) $$\cos(U, V) = \cos(V, U) = -\cos(-U, V) = -\cos(U, -V)$$
$$= \cos(-U, -V) = \cos(\perp U, \perp V).$$

The identity (N. 38, (5))

$$UV.U'V' = \begin{vmatrix} UU' & UV' \\ VU' & VV' \end{vmatrix},$$

where one divides both members by $mg\ U.mg\ V.mg\ U'.mg\ V'$ (or if one supposes the magnitudes of those vectors are equal to the unit of measure), it becomes

(10) $\sin(U, V)\sin(U', V') = \sin(U, U')\sin(V.V') - \sin(U, V')\sin(V, U').$

If in place of U' and V' one reads $\perp U'$ and $\perp V'$, one deduces

(11) $\sin(U, V)\sin(U', V') = \cos(U, U')\cos(V, V') - \cos(U, V')\cos(V, U').$

And thus one will be able to deduce many other formulas. Here we will limit ourselves to the following:

Setting $U' = \perp U$, $V' = \perp V$, one has

(12) $$\sin^2(U, V) = 1 - \cos^2(U, V).$$

Setting $U' = W$, $V' = \perp W$, one has

(13) $\sin(U, V) = \sin(U, W)\cos(V, W) - \cos(U, W)\sin(V, W).$

Putting $\perp V$ in place of V in this, one deduces

(14) $\cos(U, V) = \sin(U, W)\sin(V, W) + \cos(U, W)\cos(V, W).$

Formulas (13) and (14) express the sine and cosine of the angle of two vectors in functions of the sines and cosines that those vectors make with a third vector W; these are formulas for the addition of the angles.

We will also record from trigonometry the following formulas, which we will assume as definitions:

(15) $\tan(U, V) = \dfrac{\sin(U, V)}{\cos(U, V)}, \qquad \cot(U, V) = \dfrac{\cos(U, V)}{\sin(U, V)}.$

One has

(16) $\tan(U, V) = \dfrac{UV}{U \perp V}, \qquad \cot(U, V) = \dfrac{U \perp V}{UV}.$

42. If for reference vectors of a system of coordinates one takes a vector I equal in magnitude to the unit of measure, and the vector $J = \perp I$, then IJ will be the unit bivector, and one has

$$\perp I = J, \quad \perp J = -I, \quad IJ = I \perp I = J \perp J = 1, \quad I \perp J = 0.$$

Applying the preceding formulas, one deduces

(1) $$\perp (xI + yJ) = -yI + xJ.$$

(2) $$(xI + yJ)(x'I + y'J) = xy' - x'y.$$

(3) $$(xI + yJ) \perp (x'I + y'J) = xx' + yy'.$$

Supposing in the last that the second factor equals the first, one has

(4) $$(xI + yJ)(xI + yJ) = x^2 + y^2.$$

From this one deduces (N. 40 (7))

(5) $$mg(xI + yJ) = \sqrt{x^2 + y^2}.$$

Now (2) and (3) can also be read

(6) $$mg(xI + yJ).mg(x'I + y'J).\sin(xI + yJ, x'I + y'J) = xy' - x'y.$$

(7) $$mg(xI + yJ).mg(x'I + y'J).\cos(xI + yJ, x'I + y'J) = xx' + yy'.$$

These last formulas permit one to express the sine and cosine of the angle of two vectors as functions of the coordinates of the given vectors.

Intersections in the plane

43.   We will now return to a consideration of formations in general lying in the same plane. If α is a surface, and ω the unit surface, we will agree to write α in place of $\frac{\alpha}{\omega}$. Then, if u is the unit bivector, uA is the mass of the formation A; $uA = 0$ is the condition for A to be a vector; $\frac{A}{uA}$ represents the simple point coincident with A; $u\alpha$ the bivector of the surface α.

Theorem. *Between four formations of the first species $ABCD$ in the plane one obtains the relation*

(1) $$BCD.A - ACD.B + ABD.C - ABC.D = 0.$$

In fact, if $ABCD$ are points, and ABC is not zero, one can determine numbers x y z for which there results $D = xA + yB + zC$. Multiplying by BC one deduces $DBC = xABC$, whence $x = \frac{DBC}{ABC}$; analogously one has $y = \frac{ADC}{ABC}$, $z = \frac{ABD}{ABC}$. Substituting these values for x y z into the expression for D, and making the denominator disappear, one has demonstrated the formula. It has been demonstrated if ABC are not in a straight line; but since it is symmetric in the four points, it is also demonstrated if not all the four points are on a straight line. If this should occur, all the terms are set to zero, and (1) is again verified. Having demonstrated (1) if $ABCD$ are points, one demonstrates it when $ABCD$ represent any formations whatever with an argument analogous to that followed in N. 33.

Formula (1) can also be written as

(2) $$ABC.D = BCD.A + CAD.B + ABD.C.$$

(3) $$ACD.B - BCD.A = ABD.C - ABC.D.$$

If one multiplies (2) by a formation of the second species p, one has the identity between four formations of the first species, and one of the second:

(4) $$ABC.Dp = BCD.Ap + CAD.Bp + ABD.Cp.$$

And since, given ABC and p, it is valid whatever D may be, one deduces the relation between three formations of the first species and one of the second:

(5) $$ABC.p = Ap.BC + Bp.CA + Cp.AB.$$

If in (4) one supposes that $ABCD$ are points, and one sets $p = u$, one has $Ap = Bp = Cp = Dp = 1$, whence one recovers

(6) $$ABC = BCD + CAD + ABD.$$

44.   Definition. To the expression $AB.CD$ we shall attribute the meaning

(1) $$AB.CD = ACD.B - BCD.A.$$

Thus the expression $AB.CD$ represents a formation of the first species which is collinear with AB, since $(ACD.B - BCD.A)AB = 0$, and is collinear with CD, since $(ACD.B - BCD.A)CD = ACD.BCD - BCD.ACD = 0$. To it we will give the name the *intersection*, or *regressive product* of AB and CD. If AB is represented with a, CD with b, instead of $AB.CD$ we will also write $a.CD$, $AB.b$, $a.b$, ab.

The identity $ACD.B - BCD.A = ABD.C - ABC.D$ says that

(2) $$AB.CD = ABD.C - ABC.D.$$

If in (1) one reads p in place of CD, one has

(3) $$AB.p = Ap.B - Bp.A.$$

If in (2) in place of AB one reads p, and in place of C and D one reads A and B, one deduces

(4) $$p.AB = pB.A - pA.B.$$

Upon comparison of (3) with (4) one deduces

(5) $$pq = -qp.$$

Formula (1) says that $AB.CD$ is a function of the line CD, whence, if $CD = C'D'$, one has $AB.CD = AB.C'D'$; formula (2) says that $AB.CD$ is a function of the line AB, whence if $AB = A'B'$, one has $AB.CD = A'B'.CD$. From these one deduces

(6) $$(a = a') \cap (b = b') < (ab = a'b').$$

One has from formula (3) that $AB.(q + q') = A(q + q').B - B(q + q').A = (Aq + Aq').B - (Bq + Bq').A = (Aq.B - Bq.A) + (Aq'.B - Bq'.A) = AB.q + AB.q'$, whence, reading p in place of AB,

(7) $$p(q + q') = pq + pq'.$$

One demonstrates analogously

(8) $$(p + p')q = pq + p'q.$$

(9) $$(kp)q = p(kq) = k(pq).$$

If in (1) one supposes that two of the four points $ABCD$ coincide, one finds as particular cases

(10) $AB.AC = ABC.A, \quad AB.BC = ABC.B, \quad AC.BC = ABC.C.$

Thus the expression pq represents a formation of the first species that falls on p and on q, that is a distributive function with respect to both of its factors, and that must be multiplied by -1 when one changes the order of the factors. We will examine its meaning in some particular cases.

If p and q are two lines whose right-lines intersect at a point O, one determines the points P and Q for which one has $p = OP$, $q = OQ$. Then from (6) one has $pq = OP.OQ = OPQ.O$, whence pq represents the point of intersection of the two right-lines p and q, to which is affixed the number that measures the surface OPQ, that is the product of the magnitudes of the two lines with the sine of the interior angle. If the lines p and q are in magnitude equal to the unit of measure, the product pq represents their point of intersection to which is affixed the sine of the angle that is made; if the lines pq are in magnitude equal to the unit of magnitude and mutually orthogonal, the product pq represents their point of intersection with coefficient $+1$.

The formation of the first species pq is a vector if p and q are two parallel lines, or if one of the two factors is a bivector. Deserving special mention is the case in which one of the factors is the unit bivector u.

If $p = AB$, where A and B are simple points, one has $pu = AB.u = Au.B = Bu.A = B - A$, that is it is the vector of the line p, whence

$$pu = -up = \text{vector of the line } p.$$

The regressive product pq is zero when one of its factors is zero, or when p and q are lines lying on the same right-line, or when they are two bivectors.

45. The expression $ab.P$ represents the formation of the second species that projects P from the intersection ab.

One has the identity

(1) $ab.P = aP.b - bP.a.$

Indeed, if $b = MN$, then one has $ab = a.MN = aN.M - aM.N$, whence $ab.P = aN.MP - aM.NP$.

Now from the identity $MNP.a = Ma.NP + Na.PM + Pa.MN$ one recovers $aN.MP - aM.NP = aP.MN - MNP.a$, whence $ab.P = aP.b - bP.a$, which is the formula to be demonstrated.

Now (1) multiplied by Q gives

(2) $ab.PQ = aP.bQ - bP.aQ.$

One has the identity

(3) $ab.c = a.bc$

and the common value of these two members is indicated simply by abc.

In fact, if $c = PQ$, one has $ab.c = ab.PQ = aP.bQ - bP.aQ$; one also has $bc = b.PQ - bP.q$, whence $ab.c = bQ.aP - bP.aQ = ab.c$.

The equation $abc = 0$ says that the elements ab lie on c, that is that the three formations abc have an element of the first species in common.

The expression uab represents the mass of the formation ab, whence $uab = mg\ a.mg\ b.\sin(a, b)$.

The equation $uab = 0$ says that the right-lines a and b are parallel. If uab is not zero, $\frac{ab}{uab}$ represents the simple point of intersection of the right-lines a and b.

46.   Theorem. *If $A_1A_2A_3$ are three formations of the first species, lying in the plane, and their product is not zero, then, if A is any formation of the first species, and a one of the second, one can always determine numbers $x_1x_2x_3$, and numbers $u_1u_2u_3$ such that one has*

(1) $A = x_1A_1 + x_2A_2 + x_3A_3,$

(2) $a = u_1A_2A_3 + u_2A_3A_1 + u_3A_1A_2.$

In fact, the identities

$$A_1A_2A_3.A = AA_2A_3.A_1 + A_1AA_3.A_2 + A_1A_2A.A_3$$

and

$$A_1A_2A_3.a = A_1a.A_2A_3 + A_2a.A_3A_1 + A_3a.A_1A_2,$$

divided by $A_1A_2A_3$, which is not zero, take the places of formulas (1) and (2), where one sets

$$x_1 = \frac{AA_2A_3}{A_1A_2A_3}, \quad x_2 = \frac{A_1AA_3}{A_1A_2A_3}, \quad x_3 = \frac{A_1A_2A}{A_1A_2A_3},$$

$$u_1 = \frac{A_1a}{A_1A_2A_3}, \quad u_2 = \frac{A_2a}{A_1A_2A_3}, \quad u_3 = \frac{A_3a}{A_1A_2A_3}.$$

Definition. To the numbers $x_1x_2x_3$ and $u_1u_2u_3$, which satisfy equations (1) and (2) we will give the names the *coordinates* of A and of a, with respect to the reference elements $A_1A_2A_3$.

To simplify the expressions we will put

(5) $a_1 = A_2A_3, \quad a_2 = A_3A_1, \quad a_3 = A_1A_2.$

(6) $\theta = A_1A_2A_3.$

Then one has

(7) $A_1a_1 = A_2a_2 = A_3a_3 = \theta$

(8) $a_2a_3 = \theta A_1, \quad a_3a_1 = \theta A_2, \quad a_1a_2 = \theta A_3.$

$$a_1a_2a_3 = \theta^2.$$

One recovers, upon following the operations indicated in the first members with the rules explained, the following formulas:

(9) $(x_1A_1 + x_2A_2 + x_3A_3) + (y_1A_1 + y_2A_2 + y_3A_3)$

$$= (x_1 + y_1)A_1 + (x_2 + y_2)A_2 + (x_3 + y_3)A_3,$$

(10) $k(x_1A_1 + \ldots) = kx_1A_1 + \ldots$

(11) $(u_1a_1 + \ldots) + (v_1a_1 + \ldots) = (u_1 + v_1)a_1 + \ldots$

(12) $k(u_1a_1 + \ldots) = ku_1a_1 + \ldots$

$$(13) \quad (x_1 A_1 + x_2 A_2 + x_3 A_3)(y_1 A_1 + y_2 A_2 + y_3 A_3) = \begin{vmatrix} x_1 & x_2 & x_3 \\ y_1 & y_2 & y_3 \\ a_1 & a_2 & a_3 \end{vmatrix}$$

$$(14) \ (x_1 A_1 + x_2 A_2 + x_3 A_3)(u_1 a_1 + u_2 a_2 + u_3 a_3) = (x_1 u_1 + x_2 u_2 + x_3 u_3)\theta.$$

$$(15) \quad (u_1 a_1 + u_2 a_2 + u_3 a_3)(v_1 a_1 + v_2 a_2 + v_3 a_3) = \begin{vmatrix} u_1 & u_2 & u_3 \\ v_1 & v_2 & v_3 \\ A_1 & A_2 & A_3 \end{vmatrix} \theta.$$

(16)
$$(x_1 A_1 + x_2 A_2 + x_3 A_3)(y_1 A_1 + y_2 A_2 + y_3 A_3)(z_1 A_1 + z_2 A_2 + z_3 A_3)$$
$$= \begin{vmatrix} x_1 & x_2 & x_3 \\ y_1 & y_2 & y_3 \\ z_1 & z_2 & z_3 \end{vmatrix} \theta^2.$$

(17)
$$(u_1 a_1 + u_2 a_2 + u_3 a_3)(v_1 a_1 + v_2 a_2 + v_3 a_3)(w_1 a_1 + w_2 a_2 + w_3 a_3)$$
$$= \begin{vmatrix} u_1 & u_2 & u_3 \\ v_1 & v_2 & v_3 \\ w_1 & w_2 & w_3 \end{vmatrix} \theta^2.$$

The preceding formulas express, as functions of the coordinates of the formations given, the coordinates of every formation that one obtains from them with
the operations: the sum of two formations of the same species, the product of a
formation with a number, the progressive or regressive product of two formations.

Of the formulas

$$(18) \qquad (x_1 A_1 + x_2 A_2 + x_3 A_3)u = x_1 A_1 u + x_2 A_2 u + x_3 A_3 u,$$
$$(19) \qquad (u_1 a_1 + u_2 a_2 + u_3 a_3)u = u_1 a_1 u + u_2 a_2 u + u_3 a 3 u,$$

the first expresses the mass of a formation of the first species as a function of its
coordinates, and of the masses of the reference formations; the second, the vector
of a formation of the second species as a function of its coordinates, and of the
vectors of the reference formations. One can operate on the vectors thus obtained
with the symbol $\perp$; in this way one can express as functions of the coordinates of the
elements given, and of the elements that depend only on the reference elements,
all the functions that one obtains with the geometric operations introduced; as
particular cases, the magnitude of a line, the sine and cosine of the angle of two
lines, and so on.

These formulas are in general somewhat complicated. We therefore develop
some of them in the system of coordinates that follows.

47. We will take for reference elements a vector I equal in magnitude to the unit of measure, the vector $J =\perp I$, and a simple point O. Then the bivector IJ is the unit bivector u, and between I and J there holds the relations of N. 42. The surface $IJO = uO$ is the unit surface ω.

Every function of first species A is reducible to the form

$$(1) \qquad\qquad A = mO + xI + yJ.$$

Multiplying it by u, one has $Au = m$, whence m is the mass of the system. Thus every simple point can be cast in the form

$$(2) \qquad\qquad A = O + xI + yJ.$$

The numbers x and y are called the *Cartesian coordinates* of the point A. Every vector is expressed, as one has already seen in N. 42, by

$$(3) \qquad\qquad V = xI + yJ.$$

The identity

$$(4) \qquad (O + xI + yJ) - (O + x'I + y'J) = (x - x')I + (y - y')J$$

gives the coordinates of a vector as functions of the coordinates of its extrema. Applying formula (5) of N. 42, one deduces

$$(5) \quad mg\,|(O + xI + yJ) - (O + x'I + y'J)| = \sqrt{(x - x')^2 + (y - y')^2}$$

which gives the distance between the two points as a function of their coordinates.

Every formation of the second species is reducible to the form

$$a = pJO + qOI + ru.$$

The sum of the first two terms is a line passing through the origin, the last is a bivector.

Multiplying by $u = IJ$, and observing that $JO.u = -J$, and $OI.u = I$, one deduces

$$(7) \qquad\qquad au = qI - pJ.$$

which gives the vector of a line as a function of its coordinates. Since $mg\ a = mg\ (au)$, one deduces from the formulas of N. 42 and from the preceding that

$$(8) \qquad\qquad mg(pJO + qOI + ru) = \sqrt{p^2 + q^2}$$

One has

$$(9) \qquad\qquad (O + xI + yJ)(pJO + qOI + ru) = px + qy + r$$

which expresses the area of a triangle knowing the coordinates of the vertices and of the base. The condition for a point A with coordinates x, y to be on a right-line

a of coordinates p, q, r, which in general is expressed as $Aa = 0$, upon introducing the coordinates of the two elements becomes

(10) $$px + qy + r = 0.$$

The number $\frac{Aa}{mg\,a}$ measures the distance of the point A from the right-line a, taking the sign $+$ or $-$ according as the triangle Aa is positive or negative. Introducing the coordinates of A and a, one has

(11) $$\frac{Aa}{mg\,a} = \frac{px + qy + r}{\sqrt{p^2 + q^2}}$$

Developing the regressive product of two lines, one has

$$(pJO + qOI + ru)(p'JO + q'OI + r'u)$$
(12) $$= (pq' - p'q)O + (qr' - q'r)I + (rp' - r'p)J$$

expressed by means of the coordinates of the two lines. If $pq' - p'q$ is zero, the intersection of the two lines is a vector, and they are parallel; if it is not zero, dividing the formation obtained by $pq' - p'q$ gives the Cartesian coordinates of the point of intersection of the two lines.

Applications

48.   1. If $IJKL$ are four vectors in the plane, one calls $\frac{IK}{JK} : \frac{IL}{JL}$ the *double ratio* of the four vectors; it is not altered if one multiplies the given vectors by arbitrary numbers, and thus depends only on the directions of the four vectors. Four vectors are called *harmonic* if their double ratio equals -1, that is, if $IK.JL + JK.IL = 0$. The vectors $I, J, xI + yJ, xI - yJ$ are harmonic, whatever the numbers x and y. For the fourth harmonic after I, J, K one can take $IK.J + JK.I$.

2. The expression $U \perp V$ represents, as one has seen, the product of the magnitudes of the two vectors U and V with the cosine of the interior angle. This number is a symmetric and distributive function of the two vectors; one also calls it the *inner product* of the two vectors. Agreeing, in these exercises, to indicate with U^2 the number $U \perp U = (mg\,U)^2$, one has the following identities:

$$(U \pm V)^2 = U^2 + V^2 \pm 2U \perp V$$
$$(U + V) \perp (U - V) = U^2 - V^2.$$

If A, B, C, D are four points in the plane:

$$(A - B) \perp (C - D) + (B - C) \perp (A - D) + (C - A) \perp (B - D) = 0.$$

3. An equation containing a variable point P expresses a condition that is satisfied in general by a system of points, constituting a geometric locus; this equation is called the equation of that geometric locus.

The equation $(P - O) \perp I = 0$, where O is a fixed point, I a given vector, and P a variable point, is satisfied by all points P that lie on a right-line passing through O and normal to the vector I; thus it is the equation of this line.

The equation $(P - O) \perp I = k$, where k is a constant, represents a right-line normal to I.

An equation of the form

$$m_1(P - A_1) \perp I_1 + m_2(P - A_2) \perp I_2 + \ldots + m_n(P - A_n) \perp I_n = k$$

represents a right-line perpendicular to the vector $m_1 I_1 + m_2 I_2 + \ldots + m_n I_n$, assuming this is not zero.

4. The equation of the circle with center O and of radius can be written $mg(P - O) = r$, or $(P - O) \perp (P - O) - r^2 = 0$, or also, adopting the conventions of exercise 2, $(P - O)^2 - r^2 = 0$. The quantity $(P - O)^2 - r^2$, where P is any point whatever of the plane, one calls the *power* of the point P with respect to the circle of center O and radius r; it is positive, zero or negative, according as the point is outside, on, or inside the circle.

The locus of points having the same power with respect to the two circles of centers O and O' and radii r and r' have for an equation $(P - O)^2 - r^2 = (P - O')^2 - r'^2$, or $((P - O) + (P - O')) \perp ((P - O) - (P - O')) = r^2 - r'^2$, that is

$$\left(P - \tfrac{O+O'}{2}\right) \perp (O' - O) = \frac{1}{2}(r^2 - r'^2),$$

which represents a right-line perpendicular to OO', and which is called the *radial axis* of the two circles.

5. Every function $f(P)$ of the point P that is a sum of terms of the form $m(P - A) \perp (P - B)$, $m(P - A)^2$, $m(P - A) \perp I$, $m(P - A)I$, m, where the m are any numerical coefficients, $A, B, \ldots$ are fixed points, and the I are fixed vectors, is always reducible to the form

$$f(P) = m(P - O)^2 + n(P - O) \perp I + p,$$

where O is an arbitrary point. Indeed it suffices to substitute the difference $(P - O) - (A - O)$ for the vector $P - A$, and to order the result according to $P - O$.

If m is not zero, set $O' - O - \frac{n}{2m} \perp I$, and $k = \frac{p}{m} - \frac{n^2}{4m^2}I^2$, the function $f(P)$ can also be reduced to the form

$$f(P) = m[(P - O')2 + k].$$

The equation $f(P) = 0$, if k is negative, therefore represents a circle with center O' and radius $\sqrt{-k}$.

The equation $(P - A) \perp (P - B) = 0$ represents the locus of points P such that the right-lines PA and PB are orthogonal. This equation can also be written $\left(P - \frac{A+B}{2}\right)^2 - \frac{1}{4}(A - B)^2 = 0$, and thus this locus is a circle with center $\frac{A+B}{2}$ and radius $\frac{1}{2}mg(A - B)$.

If $A_1 \dots A_n$ are fixed points, $m_1 \dots m_n$ given numbers whose sum is not zero, and putting $m_1 A_1 + \dots + m_n A_n = (m_1 + \dots + m_n)G$, the function of the point P

$$f(P) = m_1(P - A_1)^2 + \dots + m_n(P - A_n)^2$$

is reducible to the form

$$f(P) = (m_1 + \dots + m_n)(P - G)^2 + m_1(G - A_1)^2 + \dots + m_n(G - A_n)^2.$$

Thus the locus of points for which $f(P)$ is constant, that is for which the sum of the squares of the distances from the given points, multiplied by the numbers $m_1 \dots m_n$, is constant, is a circle with center G.

The locus of points for which the ratio of the distances from two given points A and B is a constant has for an equation $\frac{mg(P-A)}{mg(P-B)} = k$, or $(P - A)^2 - k(P - B)^2 = 0$. It is therefore a circle whose center is at $\frac{A - k^2 B}{1 - k^2}$ (supposing $1 - k^2 \neq 0$).

6. If ABC are the vertices of a triangle, the identities

$$AB + BC + CA = (A - B)(A - C) = (B - C)(B - A) = (C - A)(C - B)$$

say that the area of the triangle is equal to the product of two sides with the sine of the included angle. From this one obtains that the sides in the triangle are proportional to the sines of the opposite angles.

From the identity

$$B - C = (B - A) - (C - A)$$

one recovers

$$(B - C)^2 = (B - A)^2 + (C - A)^2 - 2(B - A) \perp (C - A),$$

or

$$\begin{aligned}|mg(B - C)|^2 = |mg(B - A)|^2 + |mg(C - A)|^2 \\ - 2mg(B - A).mg(C - A).\cos(B - A, C - A)\end{aligned}$$

which expresses a side of a triangle as a function of the other two sides and the cosine of the included angle.

7. If A is a point, a a line, the right-line $A.ua$ is the parallel to a produced through A; one can also cast this into the form $Aa.u - Au.a$. The right-line $A \perp (ua)$ is the perpendicular to a through A. The expression $[A \perp (ua)]a$ represents the base of the perpendicular dropped from A on a, with mass $(mg\ a)^2$. If $mg\ a = 1$, then the preceding expression represents the base of that perpendicular.

If a and b are lines, $\frac{a}{mg\ a} \pm \frac{b}{mg\ b}$ are the bisectors of the two lines; if the magnitudes of the given lines equal the unit of measure, these bisectors are represented by $a \pm b$.

The expression $ab. \perp (uc)$ represents the perpendicular dropped from the point ab on the right-line c. One can also write it $a \perp (uc).b - b \perp (uc).a$. If the lines a, b, c are equal in magnitude to the unit of measure, one has $a \perp uc = \cos(a, c)$, $b \perp uc = \cos(b, c)$, whence the normal dropped from the point ab on c can be written $\cos(a, c).b - \cos(b, c).a$.

If a, b, c are three lines in the plane that intersect pairwise, given A, B, C as their points of intersection, one has $A = \frac{bc}{ubc}$, $B = \frac{ca}{uca}$, $C = \frac{ab}{uab}$, whence one recovers for the area of the triangle having these three right-lines for sides

$$ABC = \frac{(abc)^2}{(ubc)(uca)(uab)}.$$

8. Let ABC be three points in the plane which determine a triangle. We call a, b, c the lengths of its three sides, and we will determine the three lines a, b, c such that one has $BC = aa$, $CA = bb$, $AB = cc$. Thus a, b, c are lines equal in magnitude to the unit of measure, and lying on the sides of the triangle. We will also indicate with A the angle $(B - A, C - A)$; and analogously for B and C. Then the bisectors of the angles of the triangle are $b \pm c$, $c \pm a$, $a \pm b$; they pass, three by three, through the four points $bc \pm ca \pm ab = A \sin A \pm B \sin B \pm C \sin C$.

The perpendicular dropped from A on BC can be written $b \cos B - c \cos C$. The point of intersection of the elevations is

$$(b \cos B - c \cos C)(c \cos C - a \cos A)$$
$$= bc \cos B \cos C + ca \cos C \cos A + ab \cos A \cos B$$
$$= A \sin A \cos B \cos C + B \sin B \cos A \cos C + C \sin C \cos A \cos B;$$

dividing by the product of the cosines, the same point can be represented by

$$A \tan A + B \tan B + C \tan C.$$

The center of the circle circumscribing the triangle given is the point of intersection of the altitudes of the triangle with vertices $A + B$, $B + C$, $C + A$, in other words it is

$$(B + C) \tan A + (C + A) \tan B + (A + B) \tan C.$$

Carrying out some transformations, one can reduce this to the form[1]

$$A \tan 2A + B \tan 2B + C \tan 2C.$$

9. Indicating by $A, B \ldots , a, b, \ldots$ formations of the first and second species, one has, as has already been seen, the following identities:

$$abc.P = aP.bc + bP.ca + cP.ab.$$
$$bcd.a - acd.b + abd.c - abc.d = 0.$$
$$(BC.a)(CA.b)(AB.c) = (Ba.Cb.Ac - Ca.Ab.Bc)ABC.$$
$$abc.(BC.a)(CA.B)(AB.C) = ABC.(A.bc)(B.ca)(C.ab).$$
$$(BC.B'C')(CA.C'A')(AB.A'B') = ABC.A'B'C'.(AA'.BB'.CC').$$
$$(bc.b'c')(ca.c'a')(ab.a'b') = abc.a'b'c'.(aa'.bb'.cc').$$

Setting to zero one of the two members of this last identity, one deduces the theorem of Desargues on homological triangles. One notes that the preceding identities are valid whatever the formations introduced may be, whether points or vectors (points at infinity), lines or bivectors (right-lines at infinity), and that these elements may coincide among themselves.

10. The algebraic and geometric operations that one can carry out on numbers and geometric formations permit the expression of functions of those numbers and formations. We will say that a function of numbers and of geometric formations is *complete* {intera} if the function is expressed by means of those numbers and formations (independent variables), and by other numbers and formations (constants) with the operations:

a) addition of numbers; addition of formations of the same species.

b) multiplication of two numbers; multiplication of a number by a formation; progressive multiplication (projection) of two formations (of the first species, or one of the first and the other of the second); regressive multiplication (intersection) of two formations (of the second species); the operation $\perp$ as one carries it out on a vector; division of a bivector and of an area respectively by the unit bivector or the unit area.

A function that one expresses by means only of operations (*b*) is called a *monomial*. The *degree* {grado} of a monomial with respect to a formation is the number of times that formation appears in the expression. Every complete function is reducible to a sum of several monomials (*terms*). One calls the *degree* of a complete function the maximum of the degrees of its terms. If all the terms are of the same degree, the function is called *homogeneous*. Every monomial is a homogeneous function.

For example, the coordinates of a formation for the first or second species (N. 46, (3) and (4)) are homogeneous functions of first degree in the formation. A numerical function, homogeneous of degree n in the coordinates of a formation, is a homogeneous function of degree n in that formation, and vice-versa. The coordinates of a formation function, homogeneous of degree n in a second formation, are homogeneous functions of degree n in the second formation. The expression

$$A_0 + A_1 t + A_2 t^2 + \ldots + A_n t^n,$$

where $A_0 \ldots A_n$ are constant (independent of t) formations of the first species, represents a complete formation of the first species function of degree n in the numerical variable t; and vice versa, every complete formation of the first species function of degree n in t can be reduced to the form displayed. The expressions

$$(a + bt)(a' + b't),$$

$$(A + Bt)(A' + B't).(a + bt), \quad (A + Bt)(A' + B't).(A'' + B''t)(A''' + B'''t)$$

represent formations of the first species functions respectively of second, third and fourth degree in t, products of functions of first degree. $mg\, a$ is an incomplete function of the line a; but its square $mg^2 a = ua \perp ua$ is a complete homogeneous function of 2nd degree in a.

11. Let A, B, C be three formations of the first species. Let us put

$$(1) \qquad\qquad P = Au^2 + Buv + Cv^2.$$

Then P is a formation of the first species, a homogeneous function of second degree in the two numerical variables u and v. If in place of u and v one reads ku and kv, the formation P is multiplied by k^2, whence the position of the point P, or the direction of the vector P, depends only on the ratio of the two variables u and v.

Giving u and v the values $u_1, v_1; u_2, v_2; \ldots$, and calling $P_1, P_2, \ldots$ the corresponding formations P, one deduces

$$(2) \qquad P_1 P_2 = (u_1 v_2 - u_2 v_1)(ABu_1 u_2 + AC(u_1 v_2 + u_2 v_1) + BCv_1 v_2),$$

$$(3) \qquad P_1 P_2 P_3 = (u_1 v_2 - u_2 v_1)(u_1 v_3 - u_3 v_1)(u_2 v_3 - u_3 v_2)ABC.$$

If ABC is not zero, one deduces that three formations P corresponding to nonproportional values of u and v are never collinear. The system of positions of the point P, or of directions of the vector P, is called a *curve of* 2nd *degree.* In order that P be on the line p, one must have $Pp = 0$, that is

$$(4) \qquad\qquad Ap.u^2 + Bp.uv + Cp.v^2 = 0,$$

an equation of second degree with respect to u : v. The line p intersects the curve, or is tangent to it, or does not intersect it according as

$$(Bp)^2 - 4Ap.Cp \gtreqless 0.$$

The condition for the right-line p to be tangent to the curve is

$$(5) \qquad\qquad (Bp)^2 - 4Ap.Cp = 0.$$

The mass of P is given by

(6) $$uP = uA.u^2 + uB.uv + uCv^2.$$

This mass will be zero, and thus P a vector (point at infinity) if $uA.u^2 + uB.uv + uCv^2 = 0$, an equation of second degree in u : v; the curve contains two vectors, or one, or none (that is, it has two points at infinity, or one, or none), according as

(7) $$(uB)^2 - 4uA.uC \gtreqless 0.$$

In these three cases the curve is called respectively an *hyperbola*, a *parabola*, and an *ellipse*.

The right-line

(8) $$p = ABu^2 + 2ACuv + BCv^2$$

is the *tangent* to the curve at the point (u, v).

Multiplying (1) by BC, CA, AB, one deduces: $PBC = u^2ABC$, $PCA = uvABC$, $PAB = v^2ABC$, whence

(9) $$(PCA)^2 - (PBC)(PAB) = 0,$$

which is the condition for the element P to belong to the curve.

The simple point coincident with the formation P, assuming uP is not zero, is given by

(10) $$\frac{P}{uP} = \frac{Au^2 + Buv + Cv^2}{uA.u^2 + uB.uv + uC.v^2}.$$

12. A complete homogeneous numerical function of degree n in a formation of the first species P, supposed equal to zero, establishes a relation which, if satisfied by an element P, is also satisfied by the element kP, that is, if one element satisfies it, then so also do all elements having the same position or direction. All together, these elements are called a geometric locus of degree n.

Thus the equation $Pa = 0$, if a is a line, represents the right-line a; if a is a bivector (right-line at infinity), it is satisfied upon choosing for P any vector (point at infinity).

The equation

$$(PA.a)(PB.b)C = 0$$

represents a locus of second order (conic), whose elements have the property that if one projects from A and from B with the formations PA and PB, and intersects these with a and b, the elements so obtained are collinear with C. One sees immediately that this locus contains the elements A, B, and ab.

The equation

$$(AB.DE)(BC.EF)(CD.FA) = 0$$

expresses a relation between six elements of first species $ABCDEF$; it is of second degree in each of these elements, and is satisfied if two of the elements coincide. Suppose five of the elements are given. This is an equation of second degree in the sixth, representing a locus of second order that contains the five elements given. The preceding relation is nothing but the relation between six points of a conic expressed by the hexagram of Pascal.

The equation

$$(PA.a)(PB.b)(PC.c) = 0$$

represents a locus of third order, containing the elements A, B, C, ab, bc, ca.

A locus can also be represented by an equation inhomogeneous in P, provided one adds that P is a simple point. It can be transformed into another homogeneous equation that is meaningful whatever P may be (and, when P is a simple point, has the same meaning as the proposed) if all the terms are multiplied by appropriate powers of uP, which equals 1 when P is a simple point. Thus the circle with center on the point O and of radius r can be written $(u.OP) \perp (u.OP) - (uO)^2(uP)^2.r^2 = 0$.

Let $a_1a_2 \ldots a_n$ be lines in the plane, in magnitude equal to the unit of measure: from a point P one drops perpendiculars on the given lines, whose bases are $(P \perp ua_1).a_1$, $(P \perp ua_2).a_2$, $\ldots$; the area of the polygon having these bases for vertices is given by the number that measures the bivector:

$$(P \perp ua_1.a_1)(P \perp ua_2.a_2) + (P \perp ua_2.a_2)(P \perp ua_3.a_3)$$
$$+ \ldots + (P \perp ua_n.a_n)(P \perp ua_1.a_1),$$

a function of second degree in P. Putting this area equal to a constant k, or, for homogeneity, to $(uP)^2$k, one has an equation of second degree in P, which represents a locus of second order. One can recognize that it is a circle.

13. Setting to zero a homogeneous function of degree n in a line p, one has an equation in which, if it is satisfied by a line p, it is also satisfied by all lines kp having the same position. The system of lines that satisfy this equation is called an envelope of class n.

The envelope of right-lines p that form a triangle with two given right-lines a and b of given area k is (Exercise 7)

$$(abp)^2 = k(abu)(apu)(bpu).$$

The envelope of right-lines p for which the products of the distances from two fixed points A and B is equal to a constant k has for its equation $\frac{pA}{mg\ p} \cdot \frac{pB}{mg\ p} = k$, or

$$pA.pB = kmg^2p,$$

and thus is an envelope of second class.

The envelope of right-lines p that cut the sides a, b, c of a triangle in points, in which ascending the perpendiculars to the sides, they happen to pass through the same point, has for its equation

$$(pa. \perp ua)(pb. \perp ub)(pc. \perp uc) = 0,$$

of third class, and that contains the lines a, b, c.

14. Let there be given elements of first and second species in the plane. For simplicity in language we will, in this exercise, call a *point* an element of the first species, a *right-line* one of the second. Then, conjoining the points pairwise, and marking the intersections of the right-lines, one obtains new right-lines and new points, from which, operating in the same way, we will derive new elements that can be expressed as functions of the elements given by progressive and regressive products.

Given three points one cannot recover three arbitrary lines with this construction.

Given four points $ABCD$ (vertices of a complete quadrangle), one deduces 6 right-lines (sides of the quadrangle) $AB, \ldots$; the opposite sides intersect in three points $AB.CD, \ldots$ (diagonal points); these points united pairwise determine three right-lines (diagonal right-lines). Let one put $X = AB.CD, Y = BC.AD$, $Z = CA.BD, x = YZ, y = ZX, z = XY$; then, abstracting from the numerical coefficients, and attending only to the positions of the elements, the four given points are represented by $X \pm Y \pm Z$; the six sides by $x + y, \ldots$; the diagonal points and right-lines will be called $X, \ldots, x, \ldots$; the diagonal right-lines intersect the sides in 6 new points $X \pm Y, \ldots$; these united pairwise give the four new right-lines $x \pm y \pm z$. Continuing thus one can obtain all the points and right-lines of the form $xX + yY + zZ$ or $xx + yy + zz$, in which x, y, z are integers (or commensurables), and only these elements.

CHAPTER VII

Formations in Space

Vectors

49. In space we fix at random a volume Ω, to which we will give the name the *unit volume*; we will indicate by ω the trivector of this volume, that is that trivector such that if O is any point in space, one has $O\omega = \Omega$, and we will call it the *unit trivector*.

If A is a formation of the first species, $A = m_1 A_1 + \ldots + m_n A_n$, in which $A_1 \ldots A_n$ are points, one deduces

$$\frac{A\omega}{\Omega} = m_1 + \ldots + m_n,$$

that is $\frac{A\omega}{\Omega}$ represents the mass of the system A. The equation $A\omega = 0$ says that A is a vector. If $A\omega$ is not zero, $\frac{\Omega}{A\omega} A$ represents the simple point coincident with A. If A is a volume, $\frac{A}{\Omega} \omega$ represents the trivector of the volume.

50. We will begin by concerning ourselves with vectors and their products, that is bivectors and trivectors. If it is demonstrated that IJK are three vectors, such however that IJK is not zero, every other vector U can be cast into the form

$$(1) \qquad\qquad U = xI + yJ + zK.$$

The numbers x, y, z are called the *coordinates of the vector U* with respect to the three reference vectors I, J, K.

Multiplying the preceding equation by JK, KI, JK, one recovers

$$(2) \qquad\qquad x = \frac{UJK}{IJK}, \quad y = \frac{UKI}{IJK}, \quad z = \frac{UIJ}{IJK},$$

which express the coordinates of the vector U as ratios of trivectors.

Suppose one substitutes the values of xyz given by (2) into (1); one has, after some reductions,

$$\frac{IJK}{\omega}.U = \frac{UJK}{\omega}.I + \frac{UKI}{\omega}.J + \frac{UIJ}{\omega}.K.$$

To simplify the expression we agree to suppress the unit trivector in the denominator, understanding, from now on, that IJK is no longer the trivector product of I, J, K, but the ratio of this trivector to the unit trivector.

Now one can write the preceding formula as

(3) $IJK.U = UJK.I + UKI.J + UIJ.K,$

a relation between any four vectors.

If $U = xI + yJ + zK$, $V = x'I + y'J + z'K$, one recovers for the bivector UV the expression

(4) $UV = (yz' - y'z)JK + (zx' - z'x)KI + (xy' - x'y)IJ.$

One sees from this that every bivector u can be cast into the form

(5) $u = uJK + vKI + wIJ.$

The numbers u, v, w are called the *coordinates of the bivector u*; formula (4) gives the coordinates of the bivector UV as functions of the coordinates of its factors.

If the vector U and the bivector u are given by (1) and (5) one recovers

(6) $Uu = (xu + yv + zw)IJK.$

If x, y, z, x', y', z', x'', y'', z'' are the coordinates of the three vectors U, V, W, one recovers

(7) $UVW = \begin{vmatrix} x & y & z \\ x' & y' & z' \\ x'' & y'' & z'' \end{vmatrix} IJK.$

51.   If, as in N. 44, to the expression $UV.U'V'$ is attributed the meaning

(1) $UV.U'V' = UU'V'.V - VUV'.U,$

one has as well, by virtue of identity (3) of the preceding,

(2) $UV.U'V' = UVV'.U' - UVU'.V',$

whence $UV.U'V'$ represents a vector that lies on the two bivectors UV and $U'V'$; we can call it the *regressive product*, or *intersection*, of the two bivectors. If the bivectors are represented by u and v, their regressive product can be represented as $u.v$ or uv.

For the regressive product of bivectors there exist all the identities that hold for regressive products of lines and planes, that is to say:

(3) $UV.u = Uu.V - Vu.U.$
(4) $u.v = -v.u.$
(5) $(u = u') \cap (v = v') < (u.v = u'.v').$
(6) $(u + u').v = u.v + u'.v;$ $u.(v + v') = u.v + u.v'.$
(7) $(ku)v = u(kv) = k(uv).$
(8) $UV.UW = UVW.U.$

(9) $uv.U = uU.v - vU.u.$

(10) $uv.w = u.vw$, and their common value one can represent as uvw.

(11) $(uv = 0) =$ (the bivectors u and v have the same position {giacitura}).

(12) $(uvw = 0) =$ (the bivectors u, v, w have a common vector).

Let IJK be the three reference vectors; one sets $i = JK$, $j = KI$, $k = IJ$. If $u = ui + vj + wk$, $v = u'i + v'j + w'k$, one has

$$uv = \begin{vmatrix} u & v & w \\ u' & v' & w' \\ I & J & K \end{vmatrix} IKJ.$$

52. So far we have not yet chosen any convention for the choice of the unit volume, and of the unit trivector. We will agree to take as the *unit trivector* the trivector product of the three vectors I, J, K, equal in length to the unit of measure, orthogonal pairwise, and such that the volume $OIJK$, where O is a point, is right-handed. As a consequence the *unit volume* is the right-handed tetrahedron with base the unit surface, and of height the linear unit; its magnitude is one-sixth of the cube with side of the linear unit.

Definition. We will say that the vector U is the *index of the bivector u*, or that the bivector u is the *index of the vector U*, and we will write

$$U = |u, \quad u = |U,$$

when the direction of U and the position of u are normal to one another, the magnitudes of U and u are equal, and their senses are such that the trivector $Uu = uU$ is positive.

53. There results immediately from the definition given:

(1) $\qquad\qquad (u = u') = (|u = |u'); \quad (U = U') = (|U = |U').$

(2) $\qquad\qquad\qquad ||u = u, \quad ||U = U.$

(3) $\qquad\qquad\quad |ku = k|u, \quad |kU = k|U.$

One also has the following formulas:

(4) $\qquad\qquad\qquad |(U + V) = |U + |V.$

(5) $\qquad\qquad\qquad |(UV) = (|U).(|V).$

(6) $\qquad\qquad\qquad U|V = V|U.$

Formula (4) expresses the distributive property of the operation $|$; (5) says that the index of the bivector UV is the regressive product of the two bivectors $|U$ and

$|V$; (6) says that the trivector $U|V$, or the number that measures it, is a symmetric function of U and V.

To demonstrate these formulas, let I be a vector equal in magnitude to the unit of measure, and simultaneously normal to U and to V. Let i be the index of I; then i is a bivector lying in the plane UV, equal to the unit of measure, and of determinate sense. Now in the plane UV there is defined the operation indicated by the symbol $\perp$ acting on vectors lying in this plane, and one has

a)
$$\perp (U+V) = \perp U + \perp V, \quad b) \quad UV = (\perp U)(\perp V), \quad c) \quad U \perp V = V \perp U.$$

In addition, if X is any vector whatever lying in that plane, one has

d)
$$|X = (\perp X)I,$$

that is, the index of the vector X is the bivector product of $\perp X$ and I; indeed, the vector X and the bivector $(\perp X)I$ are orthogonal, have equal magnitude, and their product $X(\perp X)I = (mg\ X)^2 iI$ is a positive trivector.

Thus if one multiplies *a)* by I, one deduces

$$[\perp (U + V)]I = (\perp U)I + (\perp V)I,$$

and by virtue of *d)* one has (4).

To demonstrate (5), one has successively by *d)*, by (8) of N. 51, and by *c)*:

$$(|U).(|V) = [(\perp U)I].[(\perp V)I] = [(\perp U)(\perp V)I]I = [UVI]I = |(UV).$$

For (6), one multiplies *c)* by I; one has $U(\perp V)I = V(\perp U)I$, and thus, by *d)*, one has demonstrated the formula.

If, in (4), in place of U and V, one reads $|u$ and $|v$, one deduces, interchanging the two members:

$$u + v = |(|u + |v),$$

which permits one to reduce, with the operation $|$, the sum of two bivectors to that of two vectors.

Taking the indices of both of the members one has

(4')
$$|(u + v) = |u + |v.$$

With the same operation one recovers from (5)

$$uv = |(|u.|v),$$

which permits one to reduce the regressive product of two bivectors to the progressive product of two vectors. Taking the indices of both members, one has

(5')
$$|(uv) = |u.|v.$$

With the same operation one recovers from (6)

(6′) $u|v = v|u.$

If I is a vector equal in magnitude to the unit of measure, then $I|I$ is the unit trivector, that is $I|I = 1$.

If U is any vector, setting $I = \dfrac{U}{mg\ U}$, one has that I is a vector equal to the unit of measure; thus one has

$$U = (mg\ U)I, \quad |U = (mg\ U)|I, \quad U|U = (mg\ U)^2 I|I,$$

that is

(7) $U|U = (mg\ U)^2.$

Analogously one has

(7′) $u|u = (mg\ u)^2.$

The conditions for the orthogonality of two vectors U and V, of the vector U and the bivector u, and of two bivectors u and v, are respectively

$$U|V = 0, \quad U|u = 0, \quad u|v = 0.$$

One also notes the formulas

(8) $U + |u = |(u + |U), \quad u + |U = |(U + |u).$
(9) $U|u = -|(u|U), \quad u|U = -|(U|u),$

which one recovers from (4) and (5) upon reading $|u$ in place of V, and

(10) $UVW = |U.|V.|W, \quad uvw = |u.|v.|w,$

which one recovers from (6) upon reading $|(VW)$ in place of V.

54. In space, if the vectors I and J, and bivectors i and j, are in magnitude equal to the unit of measure, we put by definition

(1) $\sin(I, J) = mg(IJ), \quad \sin(I, i) = Ii, \quad \sin(i, I) = iI,$
$$\sin(i, j) = mg(ij).$$

(2) $\cos(I, J) = \sin(I, |J), \quad \cos(I, i) = \sin(I, |i),$
$$\cos(i, |I) = \sin(i, |I), \quad \cos(i, j) = \sin(i, |j).$$

These last formulas can also be written

(2′) $\cos(I, J) = I|J, \quad \cos(I, i) = mg(I|i), \quad \cos(i, I) = mg(i|I),$
$$\cos(i, j) = i|j.$$

If the vectors I, J, and the bivectors i, j are not in magnitude equal to the unit of measure, for the sines and cosines of their angles we mean the sines and cosines of the angles of the vectors and bivectors $\frac{I}{mg\,I}$, $\frac{J}{mg\,J}$, $\frac{i}{mg\,i}$, $\frac{j}{mg\,j}$, having the same directions, or positions, of the given vectors and bivectors, and of the same sense, but in magnitude equal to the unit of measure.

One deduces from the definitions given that the sine of two vectors, or of two bivectors, lies between 0 and 1, but that the sine of a vector with a bivector, or of a bivector with a vector, is a number between -1 and $+1$. The opposite holds for cosines.

Letting a be a vector or a bivector, and b be a vector or a bivector, one has

(3) $\sin(a, b) = \sin(b, a)$, $\cos(a, b) = \cos(b, a)$.

(4) $\sin(a, a) = 0$, $\cos(a, a) = 1$.

(5) $\sin(a, |a) = 1$, $\cos(a, |a) = 0$.

(6) $\sin(a, b) = \cos(a, |b) = \cos(|a, b) = \sin(|a, |b)$.

(7) $\cos(a, b) = \sin(a, |b) = \sin(|a, b) = \cos(|a, |b)$.

If a, b are two vectors, or two bivectors, one has

(8) $\sin(a, b) = \sin(-a, b) = \sin(a, -b)$.

(9) $\cos(a, b) = -\cos(-a, b) = -\cos(a, -b)$.

If instead one of a and b is a vector I and the other a bivector i, one has

(10) $\sin(I, i) = -\sin(-I, i) = -\sin(I, -i)$.

(11) $\cos(I, i) = \cos(-I, i) = \cos(I, -i)$.

One also has the formulas

(12) $mg(IJ) = mg\,I\,mg\,J \sin(I, J)$; $iI = Ii = mg\,I\,mg\,i \sin(I, i)$,

$mg(ij) = mg\,i\,mg\,j \sin(i, j)$.

(13) $I|J = mg\,I\,mg\,J \cos(I, J)$; $mg(i|I) = mg\,i\,mg\,I \cos(i, I)$,

$i|j = mg\,i\,mg\,j \cos(i, j)$.

Between the sines and cosines of the angles of the vectors and bivectors there hold relations that one obtains immediately from the identities noted between vectors and bivectors. We will limit ourselves to the following. One has

$$I J.ij = \begin{vmatrix} Ii & Ij \\ Ji & Jj \end{vmatrix}$$

where, by virtue of (12), dividing by $mg\,I\,mg\,J\,mg\,i\,mg\,j$ one deduces

(14) $\sin(I, J)\sin(i, j)\sin(IJ, ij) = \sin(I, i)\sin(J, j) - \sin(I, j)\sin(J, i)$.

If in this one sets $i = |K$, $j = |L$, upon observing that

$$ij = (|K)(|L) = |(KL),$$

one deduces that

(15)
$$\sin(I, J) \sin(K, L) \cos(IJ, KL) = \cos(I, K) \cos(J, L) - \cos(I, L) \cos(J, K).$$

Suppose, in this last, that $L = I$; one deduces

(16) $\cos(J, K) = \cos(I, J) \cos(I, K) + \sin(I, J) \sin(I, K) \cos(IJ, IK),$

which is the fundamental formula of spherical trigonometry.

55.   Take as reference vectors in space three vectors IJK equal in magnitude to the unit of measure, orthogonal pairwise, and whose trivector is positive. Then one has

$$IJK = \omega = 1.$$

If one puts

$$i = JK, \quad j = KI, \quad k = IJ,$$

one recovers

$$I = jk, \quad J = ki, \quad K = ij;$$
$$|I = i, \quad |J = j, \quad |K = k; \quad |i = I, \quad |j = J, \quad |k = K;$$
$$I|I = J|J = K|K = Ii = Jj = Kk = IJK = \omega = 1;$$
$$I|J = I|K = J|K = 0.$$

The formulas of N. 50 and 51 become, in this case,

(1) $(xI + yJ + xK)(x'I + y'J + z'K) = \begin{vmatrix} x & y & z \\ x' & y' & z' \\ i & j & k \end{vmatrix}$

(2) $(xI + yJ + zK)(x'i + y'j + z'k) = xx' + yy' + zz',$

(3) $(xi + yj + zk)(x'i + y'j + z'k) = \begin{vmatrix} x & y & z \\ x' & y' & z' \\ I & J & K \end{vmatrix}.$

One has in addition

(4) $|(xI + yJ + zK) = xi + yj + zk,$
$$|(xi + yj + zk) = xI + yJ + zK.$$

(5) $(xI + yJ + zK)|(x'I + y'J + z'K) = xx' + yy' + zz'.$

(6) $$(xI + yJ + zK)|(x'i + y'j + z'k) = \begin{vmatrix} x & y & z \\ x' & y' & z' \\ i & j & k \end{vmatrix},$$

(7) $$(xi + yj + zk)|(x'I + y'J + z'K) = \begin{vmatrix} x & y & z \\ x' & y' & z' \\ I & J & K \end{vmatrix},$$

(8) $$(xi + yj + zk)|(x'i + y'j + z'k) = xx' + yy' + zz'.$$

If in (5) and (8) one sets $x' = x$, $y' = y$, $z' = z$, taking account of identities (7) and (7') of N. 53, one has

(9) $$mg(xI + yJ + xK) = mg(xi + yj + zk) = \sqrt{x^2 + y^2 + z^2}.$$

Finally, the preceding identities, together with (12) and (13) of N. 54, permit one to recover the sine and cosine of two vectors, of a vector and a bivector, and of two bivectors as functions of their coordinates.

Intersections

56. We pass now to a consideration of formations in general in space. We will agree, in the formulas that follow, to suppress the unit volume Ω when compared with factors or with divisors. Thus, if A is any formation of the first species, $A\omega$ represents its mass; the equation $A\omega = 0$ says that the formation A is reducible to a vector; $\frac{A}{A\omega}$ represents the point coincident with the formation A.

$A\omega$ represents the trivector of the volume A.

Theorem. *Between five formations of the first species $ABCDE$ there holds the relation*

(1) $$BCDE.A - ACDE.B + ABDE.C - ABCE.D + ABCD.E = 0.$$

Thus, if $ABCDE$ are simple points, and if $ABCD$ is not zero, one can determine numbers x, y, z, t for which there results

$$E = xA + yB + zC + tD.$$

Multiplying by BCD, one has $EBCD = xABCD$, whence $x = \frac{EBCD}{ABCD}$; and analogous formulas for y, z, t; substituting these values of x, y, z, t into the expression for E, one has formula (1). This formula is demonstrated if $ABCD$ do not lie in the same plane; but since it is symmetric with respect to the five points, it is demonstrated if not all of them lie in the same plane. Should all lie in the same plane, all the terms of (1) are set to zero, and it is also true. Thus formula (1) is demonstrated if $ABCDE$ are simple points. One deduces that it is also true if those letters represent formations of the first species by an argument analogous to that employed in N. 33.

If in (1) one supposes $ABCDE$ are simple points, and one multiplies it by the unit trivector ω, since $A\omega = B\omega = \ldots = 1$, one deduces the identity (N. 18, 4th)

(2) $$BCDE - ACDE + ABDE - ABCE + ABCD = 0.$$

57. Definition. To the expression $AB.PQR$ or $PQR.AB$ we will attribute the meaning

(1) $$AB.PQR = APQR.B - BPQR.A.$$

Thus the expression $AB.PQR$ represents a formation of the first species that lies on AB and on PQR; we will call it the *intersection* or *regressive product* of the two elements AB and PQR. If AB is represented by a, and PQR by α, their product is represented by $AB.\alpha$, or $a.PQR$, or $a.\alpha$, or $a\alpha$, or αa.

From the identity (N. 56)

$$APQR.B - BPQR.A = ABQR.P + ABRP.Q + ABPQ.R$$

and definition (1) one recovers

(2) $$AB.PQR = ABQR.P + ABRP.Q + ABPQ.R.$$

One deduces from (1) that $AB.PQR$ is a function of PQR, whence if $PQR = P'Q'R'$ one has $AB.PQR = AB.P'Q'R'$.

From (2) one deduces that the same intersection is a function of AB, that is if $AB = A'B'$ one has $AB.PQR = A'B'.PQR$.

Thus one deduces that

(3) $$(a = a') \cap (\alpha = \alpha') < (a\alpha = a'\alpha').$$

If in place of AB and PQR one reads a and α, formulas (1) and (2) can also be written

(1′) $$AB.\alpha = A\alpha.B - B\alpha.A$$

(2′) $$a.PQR = aQR.P + aRP.Q + aPQ.R.$$

One also has, by virtue of (1),

$$PQ.aR = PaR.Q - QaR.P = -aQR.P - aRP.Q.$$

Summing this identity with the preceding, and in place of PQ reading b, one deduces

(4) $$a.bR + b.aR = ab.R.$$

There results immediately from (1′) and (2′) that

(5) $$a(\alpha + \alpha') = a\alpha + a\alpha', \quad (a + a')\alpha = a\alpha + a'\alpha$$

(6) $$(ka)\alpha = a(k\alpha) = k(a\alpha).$$

One has as a particular case of (1), or of (2), that

(7) $$AB.ACD = ABCD.A.$$

From the definitions there also results

(8) $$a\alpha = \alpha a.$$

We will examine the meaning of the regressive product $a\alpha$ in some particular cases. If a is a line, α a surface, and they have a point O in common, one determines the point P and the line p so that $OP = a$, $Op = \alpha$; then $a\alpha = OP.Op = OPp.O$, whence $a\alpha$ represents the point of intersection of a and α, to which one affixes the number that measures the volume $OP.p$. This number is also the number that measures the trivector of that volume, which equals $mg\ a.mg\ \alpha.\sin(a, \alpha)$.

If a and α are equal in magnitude to the unit of measure, and mutually orthogonal, $a\alpha$ represents their point of intersection with coefficient ± 1.

The expression $a\alpha$ represents a vector if a and α are a line and a parallel surface, or if a is a bivector and α a surface not containing the bivector, or if a is a line and α a trivector.

Meriting special mention is the case in which α is the unit trivector ω. Setting $a = AB$, where A and B are points, one has $a\omega = AB.\omega = A\omega.B - B\omega.A = B - A$, that is, the expression $a\omega = \omega a$ represents the vector of the line a.

The formation $a\alpha$ is zero if the right-line a is contained in the plane α, or if the bivector a is contained in the plane α, or if a is a bivector and α is a trivector.

The definition of the product $a\alpha$ given by formula (1) or (1′) is applicable only when the formation of the second species a can be equated to the product AB of two formations of the first species, that is when a is reducible to a line or to a bivector. If however $a\alpha$ is not zero, one can assume for the definition of the product $a\alpha$ formula (2′).

58. Definition. To the expression $ABC.PQR$ we will attribute the meaning

(1) $$ABC.PQR = APQR.BC + BPQR.CA + CPQR.AB.$$

Thus $ABC.PQR$ represents a formation of the second species that lies on ABC, and on PQR, as one discovers upon multiplying it by ABC and PQR. We will call it the *intersection* or *regressive product* of the two formations of the third species ABC and PQR. If those formations are represented by the letters α and β, their regressive product can be represented as $\alpha\beta$.

Substituting in place of PQR the letter π in (1), one has

(1') $ABC.\pi = A\pi.BC + B\pi.CA + C\pi.AB.$

One has as a particular case

(2) $ABC.ABD = ABCD.AB.$

One has in addition the following identities, which the reader can easily demonstrate:

(3) $\alpha\beta = -\beta\alpha.$

(4) $(\alpha = \alpha') \cap (\beta = \beta') < (\alpha\beta = \alpha'\beta').$

(5) $\alpha(\beta + \beta') = \alpha\beta + \alpha\beta', \quad (\alpha + \alpha')\beta = \alpha\beta + \alpha'\beta.$

(6) $(k\alpha)\beta = \alpha(k\beta) = k(\alpha\beta).$

If α and β are two surfaces that intersect, $\alpha\beta$ represents a line lying on the right-line of intersection of the planes α and β, and whose magnitude is equal to $mg\ \alpha.mg\ \beta.\sin(\alpha, \beta)$. If the two surfaces α and β are parallel, $\alpha\beta$ represents a bivector parallel to them. If ω is the unit trivector, then $\alpha\omega$ is the bivector of the surface α. The equation $\alpha\beta = 0$ is satisfied when α and β are surfaces whose planes coincide, or are two trivectors (lying in the plane at infinity).

59.   From the products given of regressive products in space, it follows that if A and B are two formations in space, of species s and s', if $s + s' \le 4$, the expression AB represents a progressive product, or projection, and is a formation of species $s + s'$. If however $s + s' > 4$, the expression AB represents a regressive product, or intersection, of species $s + s' - 4$. One has that both the progressive and the regressive products satisfy the properties

$$(A = A') \cap (B = B') < (AB = A'B')$$
$$A(B + B') = AB + AB', \quad (A + A')B = AB + A'B$$
$$(kA)B = A(kB) = k(AB).$$

In the progressive product one has

$$AB = (-1)^{ss'} BA,$$

and in the regressive

$$AB = (-1)^{(4-s)(4-s')} BA,$$

which one can also write as

$$AB = (-1)^{ss'} BA.$$

In the regressive product AB, given that X is an element of species $s + s' - 4$ common to A and B, and determining the elements P and Q, of species $4 - s'$ and $4 - s$, such that $XP = A$ and $XQ = B$, one has $AB = APQ.X$.

Let ABC be three formations of species s, s', s''. With them one can form the 12 products

$$AB.C, \quad BA.C, \quad C.AB, \quad C.BA; \quad AC.B, \quad CA.B, \quad B.AC, \quad B.CA;$$

$$BC.A, \quad CB.A, \quad A.BC, \quad A.CB,$$

each of which one obtains upon executing two multiplications, which can all be two progressives, or one progressive and one regressive, or all two regressives. In the first case each of the products represents a formation of species $s + s' + s''$; in the second case of species $s + s' + s'' - 4$, in the third of species $s + s' + s'' - 8$. Between the first four products are the relations $AB.C = (-1)^{ss'} BA.C = (-1)^{(s+s')s''} C.AB = (-1)^{s's+s''s+s's''} C.BA$, with analogous relations between the products of the second and the third quartet. We propose to write here the relations, where they exist, between three products of the three quartets, e.g., $AB.C, AC.B$ and $BC.A$.

One has

(1) $AB.C = A.BC = ABC$
(2) $AB.a = A.Ba = ABa.$

In these the products are progressive.

(3) $AB.\alpha = A\alpha.B - B\alpha.A$
(4) $ab.A = a.bA + b.aA$
(5) $a.A\alpha = aA.\alpha - a\alpha.A$
(6) $ab.\alpha = a.b\alpha + b.a\alpha$
(7) $\alpha\beta.A = \alpha A.\beta - \beta A.\alpha.$

In these the products are one progressive and one regressive.

(8) $\alpha\beta.a = \alpha.\beta a$,   and one indicates their common value with $\alpha\beta a$.
(9) $\alpha\beta.\gamma = \alpha.\beta\gamma$,   and one indicates their common value with $\alpha\beta\gamma$.

In the 9 formulas thus obtained, those equidistant from the extrema can be transformed into one another upon exchanging the elements of the first species with those of the third, and vice versa.

Between the products of four or more factors all other relations have already been found, deserving mention again are the following:

$$AB.\alpha\beta = \begin{vmatrix} A\alpha & A\beta \\ B\alpha & B\beta \end{vmatrix},$$

$$ABC.\alpha\beta\gamma = \begin{vmatrix} A\alpha & A\beta & A\gamma \\ B\alpha & B\beta & B\gamma \\ C\alpha & C\beta & C\gamma \end{vmatrix},$$

$$ABCD.\alpha\beta\gamma\delta = \begin{vmatrix} A\alpha & A\beta & A\gamma & A\delta \\ B\alpha & B\beta & B\gamma & B\delta \\ C\alpha & C\beta & C\gamma & C\delta \\ D\alpha & D\beta & D\gamma & D\delta \end{vmatrix}.$$

$\alpha.\beta\gamma\delta = \alpha\beta.\gamma\delta = \alpha\beta\gamma.\delta$, and one can indicate their common value with $\alpha\beta\gamma\delta$.

Coordinates

60. Let $A_1\ A_2\ A_3\ A_4$ be four formations of the first species whose product is not zero. If one puts

$$\alpha_1 = A_2A_3A_4, \quad \alpha_2 = -A_1A_3A_4, \quad \alpha_3 = A_1A_2A_4, \quad \alpha_4 = -A_1A_2A_3,$$

then one has the following

Theorem. *If A is a formation of the first species, a one of the second, α one of the third, one can always determine, and in a unique way, four numbers $x_1, \ldots, x_4$ and 6 numbers $p_{12}, \ldots, p_{34}$, and 4 numbers $u_1, \ldots, u_4$ such that one has*

(1) $$A = x_1A_1 + x_2A_2 + x_3A_3 + x_4A_4,$$

(2) $a = p_{12}A_1A_2 + p_{13}A_1A_3 + p_{14}A_1A_4 + p_{23}A_2A_3 + p_{24}A_2A_4 + p_{34}A_3A_4,$

(3) $$\alpha = u_1\alpha_1 + u_2\alpha_2 + u_3\alpha_3 + u_4\alpha_4.$$

Definition. The numbers $x_1, \ldots, x_4$ are called the *coordinates* of A; the numbers $p_{12}, \ldots, p_{34}$ the *coordinates* of a; the numbers $u_1, \ldots u_4$ the *coordinates* of α. The four formations $A_1\ A_2\ A_3\ A_4$ are called the reference formations of the first species; their products the reference formations of 2nd, 3rd, and 4th species.

61. In this number, in addition to demonstrating the preceding theorem, we will find some formulas useful for calculating with the coordinates.

The identity (N. 56)

$$A_1A_2A_3A_4.A - AA_2A_3A_4.A_1 + AA_1A_3A_4.A_2 - AA_1A_2A_4.A_3 + AA_1A_2A_3.A_4 = 0,$$

resolved with respect to A, which is possible because its coefficients are not zero, gives precisely formula (1), where one sets

$$x_1 = \frac{AA_2A_3A_4}{A_1A_2A_3A_4}, \quad x_2 = \frac{A_1AA_3A_4}{A_1A_2A_3A_4}, \ldots .$$

If $A = x_1 A_1 + \dots$ one recovers, k being a number,

(4) $$kA = kx_1 A_1 + \dots ,$$

which gives the coordinates of kA as functions of those of A. If $A = x_1 A_1 + \dots$ and $B = y_1 A_1 + \dots$, one recovers

(5) $$A + B = (x_1 + y_1)A_1 + \dots ,$$

which gives the coordinates of the sum $A + B$ as functions of the coordinates of A and B.

If $A = x_1 A_1 + \dots$ and $B = y_1 A_1 + \dots$, one recovers, developing the product AB and ordering,

(6) $$AB = (x_1 y_2 - x_2 y_1)A_1 A_2 + (x_1 y_3 - x_3 y_1)A_1 A_3 + \dots ,$$

and thus is cast into the form (2) the product of two formations of the first species; as a particular case cast into the form (2) is any line.

If $a = p_{12}A_1 A_2 + \dots , b = q_{12}A_1 A_2 + \dots$, one recovers

(7) $$ka = kp_{12}A_1 A_2 + \dots$$

(8) $$a + b = (p_{12} + q_{12})A_1 A_2 + \dots .$$

Since every formation of the second species is of the form $m_1 a_1 + m_1 a_2 + \dots$, in which $m_1, \dots$ are numbers, $a_1 a_2 \dots$ lines, and since the lines can be cast into the form (2), as can the product of lines with coefficients, and the sums of lines, one concludes that every formation of the second species can be cast into the form (2).

Formula (6) gives the coordinates of the product AB as functions of the coordinates of A and B; (7) gives the coordinates ka as functions of the coordinates of a; (8) expresses the coordinates of $a + b$ by means of those of a and b.

If $A = x_1 A_1 + \dots$, $B = y_1 A_1 + \dots$, $C = z_1 A_1 + \dots$, one recovers, upon developing the product ABC,

(9) $$ABC = \begin{vmatrix} x_1 & x_2 & x_3 \\ y_1 & y_2 & y_3 \\ z_1 & z_2 & z_3 \end{vmatrix} \alpha_1 + \dots$$

If $A = x_1 A_1 + \dots$, $a = p_{12}A_1 A_2 + \dots$, one recovers

(10) $$Aa = aA = (x_2 p_{34} - x_3 p_{24} + x_4 p_{23})\alpha_1 + \dots$$

If $\alpha = u_1 \alpha_1 + \dots$, $\beta = v_1 \alpha_1 + \dots$, one recovers

(11) $$k\alpha = ku_1 \alpha_1 + \dots$$

(12) $$\alpha + \beta = (u_1 + v_1)\alpha_1 + \dots$$

Formulas (10), (11), and (12) say that one can cast triangular surfaces into the form (3), and also the product of these surfaces with numbers, and the sum of such products; thus every formation of the third species is reducible to the form (3).

Formulas (9)-(12) give the coordinates of ABC, $A\alpha$, $k\alpha$, $\alpha + \beta$ as functions of the coordinates of A, B, C, of A and a, of α and β.

Thus is demonstrated the existence of numbers x, p, u that satisfy conditions (1), (2), (3). It remains to be demonstrated that these numbers are completely determinate. For this one multiplies (1) by $A_2A_3A_4, \ldots$; (2) by $A_3A_4, \ldots$; (3) by $A_1, \ldots$; dividing by $A_1A_2A_3A_4$, one recovers

(13)
$$x_1 = \frac{AA_2A_3A_4}{A_1A_2A_3A_4}, \ldots; \qquad p_{12} = \frac{aA_3A_4}{A_1A_2A_3A_4}, \ldots; \qquad u_1 = \frac{A_1\alpha}{A_1A_2A_3A_4}, \ldots$$

which express in a unique way the coordinates of A, a, α.

62. The following formulas are also to be noted:

If $A = x_1A_1 + \ldots$, $B = y_1A_1 + \ldots$, $C = z_1A_1 + \ldots$, $D = t_1A_1 + \ldots$, one recovers

(14)
$$ABCD = \begin{vmatrix} x_1 & x_2 & x_3 & x_4 \\ y_1 & y_2 & y_3 & y_4 \\ z_1 & z_2 & z_3 & z_4 \\ t_1 & t_2 & t_3 & t_4 \end{vmatrix} A_1A_2A_3A_4.$$

If $a = p_{12}A_1A_2 + \ldots$, $b = q_{12}A_1A_2 + \ldots$, one recovers

(15) $ab = (p_{12}q_{34} - p_{13}q_{24} + p_{14}q_{23} + p_{23}q_{14} - p_{24}q_{13} + p_{34}q_{12})A_1A_2A_3A_4$.

As a particular case,

(16) $\qquad\qquad aa = 2(p_{12}q_{34} - p_{13}q_{24} + p_{14}q_{23})A_1A_2A_3A_4$.

If $A = x_1A_1 + \ldots$, $a = u_1a_1 + \ldots$, one has

(17) $\qquad\qquad A\alpha = (x_1u_1 + x_2u_2 + x_3u_3 + x_4u_4)A_1A_2A_3A_4$.

Setting, for simplicity in the expressions, $A_1A_2A_3A_4 = \Delta$, one has the following regressive products

$$A_1A_2.\alpha_1 = \Delta A_2, \quad A_1A_2.\alpha_2 = \Delta A_1, \quad A_1A_2.\alpha_3 = 0, \quad A_1A_2.\alpha_4 = 0, \text{ etc.}$$

Thus, developing the regressive product on the left, one has

(18) $\quad (p_{12}A_1A_2 + \ldots)(u_1\alpha_1 + \ldots) = (p_{12}u_2 - p_{13}u_3 + p_{14}u_4)\Delta A_1 + \ldots$

One also has $\alpha_1\alpha_2 = \Delta A_3A_4$, etc. Thus, developing the regressive product on the left, one has

(19) $\quad (u_1\alpha_1 + \ldots)(v_1\alpha_1 + \ldots) = (u_1v_2 - u_2v_1)\Delta A_3A_4 + \ldots$

From the formulas $\alpha_1\alpha_2\alpha_3 = \Delta^2 A_1$, and analogously, and from $\alpha_1\alpha_2\alpha_3\alpha_4 = \Delta^3$, one recovers

$$(20) \quad (u_1\alpha_1 + \ldots)(v_1\alpha_1 + \ldots)(w_1\alpha_1 + \ldots) = \begin{vmatrix} u_2 & u_3 & u_4 \\ v_2 & v_3 & v_4 \\ w_2 & w_3 & w_4 \end{vmatrix} \Delta^2 A_1 + \ldots$$

$$(21) \qquad (u_1\alpha_1 + \ldots)(v_1\alpha_1 + \ldots)(w_1\alpha_1 + \ldots)(t_1\alpha_1 + \ldots)$$

$$= \begin{vmatrix} u_1 & u_2 & u_3 & u_4 \\ v_1 & v_2 & v_3 & v_4 \\ w_1 & w_2 & w_3 & w_4 \\ t_1 & t_2 & t_3 & t_4 \end{vmatrix} \Delta^3.$$

If ω is the unit trivector, one has

$$(x_1A_1 + \ldots)\omega = x_1A_1\omega + \ldots$$
$$(p_{12}A_1A_2 + \ldots)\omega = p_{12}A_1A_2\omega + \ldots$$
$$(u_1a_1 + \ldots)\omega = u_1a_1\omega + \ldots$$

which express the mass of a formation of the first species, the vector of one of the second, the bivector of one of the third, as functions of the coordinates of the formations, and of the mass, or vectors and bivectors of the reference formations.

Applying the operation | to the vectors and bivectors, one deduces that every function of several formations obtained upon carrying out on them the operations sum, multiplication by numbers, progressive or regressive multiplication, and the operation |, can be expressed by means of the coordinates of the given formations, and elements that depend only on the reference elements. Thus one can calculate the magnitude of a line or surface, the trigonometric functions of the angle of two lines, or of a line and a surface, or of two surfaces by means of the coordinates of those elements and numbers that depend on the system of coordinates.

63.   We will develop the preceding formulas in the special case in which one takes as reference elements a point O and three vectors IJK, equal in magnitude to the unit of measure, orthogonal pairwise, and whose trivector is positive. Then the trivector IJK will be the unit trivector: $IJK = \omega$, and the volume $OIJK = O\omega$ will be the unit volume Ω.

Every formation of the first species will be reducible to the form

$$(1) \qquad\qquad A = mO + xI + yJ + zK;$$

multiplying by ω one deduces $A\omega = m$, whence m is the mass of the system. If $m = 1$, A is a simple point, whence every simple point can be reduced to the form

$$(2) \qquad\qquad O + xI + yJ + zK;$$

the numbers x, y, z are called the orthogonal *Cartesian coordinates* of the point.
If m = 0, the formation represents a vector, which has the form (N. 50)

$$(3) \qquad\qquad xI + yJ + zK.$$

The identity

$$(4) \quad (O + xI + yJ + zK) - (O + x'I + y'J + z'K) = (x-x')I + (y-y')J + (z-z')K$$

gives the coordinates of a vector as functions of the coordinates of its extrema.
Thus from formula (9) of N. 55, one recovers the distance between two points as
functions of their coordinates.

If, as in N. 55, one puts

$$i = JK, \quad j = KI, \quad k = IJ,$$

every formation of the second species is reducible to the form

$$(5) \qquad\qquad a = pOI + qOJ + rOK + p'i + q'j + r'k.$$

Multiplying by ω one deduces

$$(6) \qquad\qquad a\omega = pI + qJ + rK,$$

which gives the coordinates of the vector of a as a function of the coordinates of a.
One also has

$$(7) \qquad\qquad aa = 2(pp' + qq' + rr'),$$

and the annuling of aa is the condition for a to be reducible to a line or to a bivector.

Every function of the third species is reducible to the form

$$(8) \qquad\qquad \alpha = sOi + tOj + uOK + v\omega.$$

Multiplying by ω one deduces

$$(9) \qquad\qquad \alpha\omega = xi + yj + zk,$$

which gives the coordinates of the bivector of α as a function of the coordinates
of α.

Now, since the magnitude of a line or of a surface, and the angle of two lines, or
of a line and a surface, or of two surfaces are equal to the magnitudes of the lines,
or to the bivectors of the surfaces, and the angles between the vectors or bivectors
of the lines or surfaces, the formulas of N. 55 immediately permit one to express
those magnitudes, and the sines and cosines of those angles as functions of the
coordinates of the elements given.

Applications

64. 1. If I, J, K are three vectors equal in magnitude to the unit of measure, one puts for the definition of the first member

(1) $$\sin(I, J, K) = IJK.$$

If I, J, K are any three vectors, one puts by definition

(2) $$\sin(I, J, K) = \sin\left(\frac{I}{mg\ I}, \frac{J}{mg\ J}, \frac{K}{mg\ K}\right).$$

If a, b, c are three right-lines of determinate sense passing through the same point, which form a *tricorn* {trispigolo}, one defines for the *sine of the tricorn* the sine of the three vectors parallel and of the same sense as the three given right-lines.

One recovers from (2)

(3) $$IJK = mg\ I.mg\ J.mg\ K.\sin(I, J, K).$$

Supposing I, J, K equal in magnitude to the unit of measure,

$$IJK = I.JK = J.KI = K.IJ,$$

which one can also write

$$\sin(I, J, K) = \sin(J, K)\sin(I, JK) = \sin(K, I)\sin(J, KI)$$
(4) $$= \sin(I, J)\sin(K.IJ).$$

whence the sine of a tricorn is equal to the product of the sine of one face with the sine of the angle of that face with the opposite edge {spigolo}.

The formula $IJ.IK = IJK.I$, upon taking the absolute values of both its members, becomes

(5) $$\sin(I, J)\sin(I, K)\sin(IJ, IK) = mg\ IJK,$$

whence the sine of a tricorn, in absolute value, is equal to the product of the sines of two faces with the sine of the angle that they make. One recovers from the same formula

$$\frac{\sin(IJ, IK)}{\sin(J, K)} = \frac{\sin(JK, JI)}{\sin(K, I)} = \frac{\sin(KI, KJ)}{\sin(I, J)}$$
(6) $$= \frac{mg\ IJK}{\sin(J, K)\sin(K, I)\sin(I, J)},$$

which expresses the proportionality between the sines of the dihedra and the sines of the opposite faces.

2. Let i, j, k be three bivectors equal in magnitude to the unit of measure. One puts by definition

(1) $$\sin(i, j, k) = ijk.$$

If i, j, k are any three bivectors, one puts

(2) $$\sin(i, j, k) = \sin\left(\frac{i}{mg\ i}, \frac{j}{mg\ j}, \frac{k}{mg\ k}\right).$$

One deduces from this last

(3) $$ijk = mg\ img\ jmg\ k\sin(i, j, k).$$

If three planes, which are determinate in position and sense, pass through the same point, and form a *trihedron*, one defines for the sine of this trihedron the sine of the three bivectors having the same positions and senses as the planes given. One demonstrates, in a way analogous to that of the preceding, that the sine of a trihedron is equal to the product of the sine of a dihedron with the sine of the angle which that edge makes with the opposite face, and that, in absolute value, equals the product of the sine of the two dihedrons with the sine of the face between them.

The formula $IJ.IK.JK = (IJK)^2$ can be read

(4) $$\sin(I, J)\sin(I, K)\sin(J, K)\sin(IJ, IK, JK) = \sin^2(I, J, K).$$

One also has

(5) $$\sin(i, j, k) = \sin(|i, |j, |k).$$

and from formulas (4) and (5) one can deduce two other formulas upon exchanging the vectors with the bivectors.

3. Let IJK be three vectors, equal in magnitude to the unit of measure. Let i, j, k be three bivectors having the positions and senses of JK, KI, IJ, and equal in magnitude to the unit of measure. Between them there hold the relations

$$JK = i\sin(J, K), \quad KI = j\sin(K, I), \quad IJ = k\sin(I, J).$$
$$jk = I\sin(j, k), \quad ki = J\sin(k, i), \quad ij = K\sin(i, j).$$

Let O be a point fixed at random; if I is any vector, by the expression 'the right-line I' we will mean the right-line passing through O and parallel to I, that is the right-line OI; and if i is any bivector, by the expression 'the plane i' we will mean the plane passing through O and parallel to i, that is the plane Oi. Then the right-lines I, J, K are the edges of a *tricorn* or *trihedron*, whose faces are the planes i, j, k.

The right-lines $I \pm J, I \pm K, J \pm K$ are the six bisectors of the faces of the trihedron. They lie, three by three, in the four planes $JK \pm KI \pm IJ =$

$i \sin(J, K) \pm j \sin(K, I) \pm k \sin(I, J)$. These planes make angles equal to the three edges of the trihedron.

The planes $IJ \pm IK, \ldots$ are the six planes passing through an edge of the trihedron, and through a bisector of the opposite face. They pass, three by three, through the four right-lines $I \pm J \pm K$.

The plane bisectors of the dihedra of the trihedron are $i \pm j, i \pm k, j \pm k$. They pass, three by three, through the four right-lines $I \sin(j, k) \pm J \sin(k, i) \pm K \sin(i, j)$: these right-lines make equal angles with the faces of the trihedron. The plane bisector $i - j$ intersects the opposite face k along the right-line $k(i - j) = ki - kj = J \sin(k, i) + I \sin(k, j)$; these right-lines intersect, three by three, in the four planes $i \pm j \pm k$.

The planes $I|i, J|j, K|k$ are planes passing through an edge and perpendicular to the opposite face; they intersect along the right-line

$$I \tan(j, k) + J \tan(k, i) + K \tan(i, j).$$

The right-lines $i|I, j|J, k|K$ are the right-lines contained in a face of the trihedron and perpendicular to the opposite edge. They lie in the same plane

$$i \tan(J, K) + j \tan(K, I) + k \tan(I, J).$$

This plane is perpendicular to the right-line last considered.

4. Let A, a, α, A be formations of $1^{\text{st}}, 2^{\text{nd}}, 3^{\text{rd}}, 4^{\text{th}}$ species. Recollecting the results previously obtained, one has:

(1) $A\omega =$ mass of the formation A.
(2) $\frac{A}{A\omega} =$ simple point coincident with A.
(3) $(A\omega = 0) =$ the formation A is reducible to a vector (point at infinity).
(4) $(aa = 0) \cap -(a\omega = 0) =$ the formation a is reducible to a line.
(5) $(a\omega = 0) =$ the formation a is reducible to a bivector (right-line at infinity).
(6) $a\omega =$ vector of the formation a.
(7) $\alpha\omega =$ bivector of the surface α.
(8) $(\alpha\omega = 0) =$ the formation α is reducible to a trivector (plane at infinity).
(9) $\mathrm{A}\omega =$ trivector of the volume A.

5. Let $A, B, \ldots$ be simple points, $a, b, \ldots$ lines, $\alpha, \beta, \ldots$ surfaces. One has:

(1) Right-line $AB =$ right-line that connects the points A and B.
(2) $(AB = 0) =$ the points A and B coincide.
(3) Plane $(A + B)|(A - B) =$ locus of points equidistant from A and B.
(4) Plane $Aa =$ plane passing through A and a.
(5) $(Aa = 0) =$ the point A lies on the right-line a.
(6) $\frac{mg\, Aa}{mg\, a} =$ distance of the point A from the right-line a.
(7) Right-line $A.a\omega =$ right-line passing through A and parallel to a.

(8)   Plane $A|(a\omega)$ = plane passing through A and normal to a.

(9)   Right-line $Aa.A|(a\omega)$ = perpendicular dropped from A on a.

(10) $A|(a\omega).a$ = base of the perpendicular dropped from A on a with mass mg^2a.

(11) $(A\alpha = 0)$ = the point A lies in the plane α.

(12) $\frac{A\alpha}{mg\,\alpha}$ = distance of the point A from the plane α, taken with the sign of the volume $A\alpha$.

(13) Plane $A.\alpha\omega$ = plane passing through A and parallel to α.

(14) Right-line $A|(\alpha\omega)$ = perpendicular dropped from A on α.

(15) $A|(\alpha\omega).\alpha$ = base of the preceding perpendicular with mass $mg^2\alpha$.

(16) $(ab = 0)$ = the right-lines a and b lie in the same plane.

(17) $(a\omega.b\omega = 0)$ = the right-lines a and b are parallel.

(18) $(a\omega.|b\omega = 0)$ = the right-lines a and b are perpendicular.

(19) $mg(a\omega.b\omega) = mg\,a.mg\,b\sin(a, b)$.

(20) $(a\omega)|(b\omega) = mg\,a.mg\,b\cos(a, b)$.

(21) Plane $a.b\omega$ = plane passing through a and parallel to b.

(22) Right-line $[a|(a\omega.b\omega)][b|(a\omega.b\omega)]$ = perpendicular common to the right-lines a and b.

(23) $a[b|(a\omega.b\omega)]$ = point of intersection of the preceding right-line with the right-line a, with mass $mg^2(a\omega.b\omega)$.

(24) $\frac{ab}{mg(a\omega.b\omega)}$ = minimum distance of the two right-lines taken with the sign of the volume ab.

(25) $\frac{ab}{mg\,a\,mg\,b}$ = (following Staudt) the moment of the two right-lines a and b.

(26) $a\alpha$ = point of intersection of the right-line a with the plane α, with mass $a\alpha\omega = mg\,amg\,\alpha.\sin(a, \alpha)$.

(27) $(a\alpha\omega = 0)$ = the right-line a is parallel to the plane α.

(28) $(a\alpha = 0)$ = the right-line a lies in the plane α.

(29) Plane $a|(\alpha\omega)$ = plane passing through a and perpendicular to α.

(30) $(a|(\alpha\omega) = 0)$ = the right-line a is perpendicular to the plane α.

(31) Right-line $[a|(\alpha\omega)]\alpha$ = orthogonal projection of the right-line a on the plane α.

(32) $\sin(a, \alpha) = \frac{a\alpha\omega}{mg\,a\,mg\,\alpha}$, $\cos|(a, \alpha) = \frac{mg(a|\alpha\omega)}{mg\,a\,mg\,\alpha}$.

(33) Right-line $\alpha\beta$ = intersection of the planes α and β.

(34) $(\alpha\beta = 0)$ = the planes α and β coincide.

(35) $(\alpha\beta\omega = 0)$ = the planes α and β are parallel.

(36) $[(\omega\alpha)|(\omega\beta) = 0]$ = the planes α and β are perpendicular.

(37) $\sin(\alpha, \beta) = \frac{mg(\alpha\beta\omega)}{mg\,\alpha\,mg\,\beta}$, $\cos(\alpha, \beta) = \frac{(\omega\alpha)|(\omega\beta)}{mg\,\alpha\,mg\,\beta}$.

(38) Plane $\frac{\alpha}{mg\,\alpha} + \frac{\beta}{mg\,\beta}$ = plane bisector of the planes α and β.

(39) $\frac{\alpha\beta\gamma}{\alpha\beta\gamma\omega}$ = point of intersection of the three planes α, β, γ.

(40) $(\alpha\beta\gamma = 0)$ = the planes α, β, γ pass through the same right-line, or are parallel to one another.

(41) $(\alpha\beta\gamma\omega = 0)$ = the planes α, β, γ are parallel to the same right-line.

(42) $(\alpha\beta\gamma\delta = 0)$ = the planes $\alpha\beta\gamma\delta$ pass through the same point, or are parallel to the same right-line.

(43) $\dfrac{(\alpha\beta\gamma\delta)^2}{\alpha\beta\gamma\omega.\alpha\beta\delta\omega.\alpha\gamma\delta\omega.\beta\gamma\delta\omega}$ = volume of the tetrahedron bounded by the four planes α, β, γ, δ.

6. Let $ABCD$ be the vertices of a tetrahedron. The identities

$$ABCD = A.BCD = AB.CD = A(B - A)(C - A)(D - A)$$

say that the volume of a tetrahedron is equal to the product of a base with the corresponding altitude, or to the product of the magnitudes of the two opposite edges with their moment, or to the product of the magnitudes of three edges concurrent in a vertex with the sine of the corresponding tricorn.

The identity $ABC.ABD = AB.ABCD$, where one takes the absolute value of both members, becomes

$$mg\ ABC.mg\ ABD.\sin(ABC, ABD) = mg\ AB.mg\ ABCD,$$

which says that a tetrahedron, in absolute value, is equal to the product of two faces with the sine of the included angle, divided by the corresponding edge.

The identity $ABC.ABD.ACD = A.ABCD^2$, where one takes the masses of both members, says that the square of the volume of a tetrahedron is equal to the product of three faces with the sine of the trihedron that they form.

The plane $\frac{ABC}{mg\ ABC} + \frac{ABD}{mg\ ABD}$ is the plane bisector of the dihedron AB. It intersects the opposite edge CD in the point

$$\left(\frac{ABC}{mg\ ABC} + \frac{ABD}{mg\ ABD}\right)CD = ABCD\left(\frac{C}{mg\ ABC} + \frac{D}{mg\ ABD}\right),$$

from which one recovers immediately that it divides the edge CD into parts directly proportional to the faces that terminate it.

The 12 plane bisectors of the dihera of the trihedron pass through the 8 points

$$A\ mg\ BCD \pm B\ mg\ ACD \pm C\ mg\ ABD \pm D\ mg\ ABC,$$

which are the centers of the spheres tangent to the faces of the tetrahedron.

The right-line $\frac{AB}{mg\ AB} + \frac{AC}{mg\ AC} + \frac{AD}{mg\ AD}$ is the right-line through which passes the planes that connect the edges of the trihedron in A with the internal bisectors of the opposite faces. In order that it intersect the correspondent of the trihedron in B, it is necessary that their progressive product be zero. Developing the calculation one has as conditions

$$mg\ AC.mg\ BD = mg\ AD.mg\ BC.$$

The formation

$$\begin{vmatrix} 0 & A & B & C & D \\ 1 & 0 & mg^2 AB & mg^2 AC & mg^2 AD \\ 1 & mg^2 BA & 0 & mg^2 BC & mg^2 BD \\ 1 & mg^2 CA & mg^2 CB & 0 & mg^2 CD \\ 1 & mg^2 DA & mg^2 DB & mg^2 DC & 0 \end{vmatrix}$$

represents the center of the circumscribed sphere of the tetrahedron with mass 8 times the square of the tetrahedron.

7. We will say that a number or a geometric formation is a complete function of numbers and geometric formations when it can be obtained from independent variables and constants with the operations:

a) Sum of two numbers, or of two formations of the same species.

b) Multiplication of two numbers; progressive multiplication of two formations (of species 1 and 1, or 1 and 2, or 1 and 3, or 2 and 2); regressive multiplication of two formations (of the second and third species, or between two of the third species); the operation | carried out on a vector or bivector; division of a trivector by the unit trivector, or of a volume by the unit volume.

A function is called a *monomial* if it can be obtained upon carrying out only the distributive operations b). One calls the *degree of a monomial* with respect to a formation the number of times that this formation appears as a factor in the monomial. Every complete function is reducible to the sum of several monomials. One calls the degree of a complete function the maximum of the degrees of its monomials. If all the monomials that comprise a complete function are of the same degree, the function is called *homogeneous*. Every monomial is a homogeneous function. If a formation Y is a complete function, homogeneous of degree n of a formation X, the coordinates of Y are complete functions homogeneous of degree n of the coordinates of X.

The quantities $mg^2 a$ and $mg^2 \alpha$, where a and α are a line and a surface, are complete functions of second degree of a and α.

8. The points P in space for which a numerical function of P is zero form, in general, a surface. If the function is complete of degree n, one says this surface is of order n.

Thus the equation $mg\ OP = r$, or $(P - O)(P - O) - r^2 = 0$ represents a sphere with center O and of radius r. The first member of the second equation is a complete function of second degree of P.

The equation of 2nd degree $[(P - O)|I]^2 - kmg^2(P - O) = 0$, where O is a point, I a vector, and k a numerical constant, represents a cone of revolution having for vertex O, and whose axis has the direction of I.

The equation $mg\ Pa = r$, which one can put into the form, of second degree in P, $mg^2 Pa = r^2$, where a represents a line, in magnitude equal to the unit of measure, and r is a numerical constant, represents the cylinder of revolution having for axis the right-line a, and for radius r.

The equation of the locus of points, for which the ratio of the distances of two given right-lines is constant, can be cast into the form $\frac{mg\ Pa}{mg\ Pb} = k$, where a and b are lines, in magnitude equal to the unit of measure, lying on the given right-lines, and k is the given constant ratio. The same equation can be written

$$mg^2 Pa - k^2 mg^2 Pb = 0,$$

and the first member is a function of second degree in P.

9. A complete numerical function, homogeneous of degree n of a formation of the first species P, when set to zero, expresses a condition such that if a formation P satisfies it, all formations of the form mP, having the same position, or direction, also satisfy it. The ensemble of these formations is called a geometric locus of order n.

Thus the equation of a sphere $(P - O)|(P - O) - r^2 = 0$, where O is a simple point and r a number, is meaningful only when P is a simple point. It can also be written

$$(OP.\omega)|(OP.\omega) - (P\omega)^2 r^2 = 0,$$

and the first member is a function homogeneous of second degree in P, having meaning whatever may be the formation of the first species P.

The equation $Pa.Pb.c = 0$, where abc are lines, represents the surface of second order containing the right-lines a, b, c.

The locus of points P such that the 4 bases of the perpendiculars dropped from P on the faces of a tetrahedron $ABCD$ lying in a plane has for its equation

$$\frac{mg^2 ABC}{PABC} - \frac{mg^2 ABD}{PABD} + \frac{mg^2 ACD}{PACD} - \frac{mg^2 BCD}{PBCD} = 0.$$

Making the denominators disappear, this becomes of third degree in P. The edges of the tetrahedron belong to the locus.

The equation

$$(PA.\alpha)(PB.\beta)c = 0$$

represents a surface of second order. It can also be written

$$P\alpha.P\beta.ABc + P\alpha.B\beta.PAc + A\alpha.P\beta.BPc = 0.$$

This surface contains the right-line $\alpha\beta$, $Bc.\alpha$, $Ac.\beta$, etc.

The equation $(Pa.\omega)|(Pb.\omega) = 0$, where a and b are lines, represents the surface of second order of loci of the points P such that the planes Pa and Pb are orthogonal. It contains the right-lines a and b.

10. A complete numerical function homogeneous of degree n of a formation of the second species p, when set to zero, is first a condition such that if a line p satisfies it, all lines mp that are on the same right-line p also satisfy it. The right-line that contains lines p that satisfy that condition form a system called a *complex of degree* n.

Let a be a formation of the second species. The equation $ap = 0$ represents a complex of first degree. If a is reducible to a line, that complex is composed of the ensemble of right-lines that intersect a, or are parallel to it. If a is reducible to a bivector, the complex contains all the right-lines parallel to that bivector. Whatever a may be, the right-lines of the complex passing through the point P lie in the plane Pa, and the right-lines of the complex lying in the plane π pass thrugh the point πa. If the point P describes a right-line p, the plane Pa passes through the right-line $2ap.a - aa.p$.

The equation $(pA.\omega)|(pB.\omega) = 0$, where A and B are points, represents the complex of second order whose right-lines p are such that the planes pA and pB are orthogonal.

11. A complete numerical function homogeneous of degree n of a formation of the third species π, when set to zero, expresses a condition such that, if a surface π satisfies it, every surface mπ contained in the same plane π also satisfies it. The ensemble of these planes π is called an envelope of class n.

The envelope of planes, for which the sum of the distances from the points $A, B, C, \ldots$ is constant, has an equation that one can write

$$[(A + B + \ldots)\pi]^2 = k^2 mg^2 \pi,$$

of second degree in π.

The envelope of planes, for which the sum of the squares of the distances from the points $A, B, \ldots$ is constant, has for its equation

$$(A\pi)^2 + (B\pi)^2 + \ldots = k^2 mg^2 \pi,$$

also of second degree.

The envelope of planes π that, with three given planes $\alpha\beta\gamma$ make a tetrahedron of given volume k, has for its equation

$$(\alpha\beta\gamma\pi)^3 = (\alpha\beta\gamma\omega)(\alpha\beta\pi\omega)(\alpha\gamma\pi\omega)(\beta\gamma\pi\omega)k,$$

of third degree in π.

The equation $\pi a . \pi b . c = 0$ represents an envelope of second degree. To it belong all the planes tangent to the surface of the equation $Pa.Pb.c = 0$.

CHAPTER VIII

Derivatives

65. Geometric formations, like numbers, can be constants or variables. We will suppose known the definitions and theorems that refer to numerical variables, and posit the following definitions for the limits of geometric formations.

Definition. We will say that the variable volume A has for its limit the fixed volume A_0, if, Ω being the unit volume, the number $\frac{A}{\Omega}$ has for its limit the number $\frac{A_0}{\Omega}$.

We will say that the variable formation of the $\begin{pmatrix} 1^{st} \\ 2^{nd} \\ 3^{rd} \end{pmatrix}$ species $\begin{pmatrix} S \\ s \\ \sigma \end{pmatrix}$ has for its limit the fixed formation $\begin{pmatrix} S_0 \\ s_0 \\ \sigma_0 \end{pmatrix}$, if, however one takes the $\begin{pmatrix} \text{surface } \pi \\ \text{line } p \\ \text{point } P \end{pmatrix}$, one has $\begin{pmatrix} \lim S\pi = S_0\pi \\ \lim sp = s_0 p \\ \lim \sigma P = \sigma_0 P \end{pmatrix}$.

One immediately reduces the propositions that follow to theorems on limits of numbers:

1. *If A and B are two formations of the second species, one has*

$$(\lim A = A_0) \cap (\lim B = B0) < (\lim(A + B) = A_0 + B_0).$$

2. $(\lim A = A_0) \cap (\lim x = x_0) < (\lim xA = x_0 A_0).$

3. $(\lim A = A_0) \cap (\lim x = x_0) \cap -(x = 0) < \left(\lim \frac{A}{x} = \frac{A_0}{x}\right).$

4. *If AB is the progressive or regressive product of the two formations A and B, one has*

$$(\lim A = A_0) \cap (\lim B = B_0) < (\lim AB = A_0 B_0).$$

5. $(\lim A = A_0) =$ *the coordinates of A have for limits the coordinates of A_0.*

6. *If I is a vector in the plane, one has*

$$(\lim I = I_0) = (\lim \perp I = \perp I0).$$

7. *If I is a vector or bivector in space, one has*

$$(\lim I = I_0) = (\lim |I = |I_0).$$

Applying the preceding propositions, one can determine the limits of more complicated expressions. Thus, if a is a formation of the second species, one has $(\lim a = a_0) < (\lim \omega a = \omega a_0) < (\lim |\omega a = |\omega a_0) < (\lim \omega a | \omega a = \omega a_0 | \omega a_0) < (\lim mg^2 a = \lim mg^2 a_0) < (\lim mg\ a = mg\ a_0)$, by virtue of 4, 7, 4, and the numerical identities.

8. In an analogous way one has, if A is a formation for which a magnitude is defined,

$$(\lim A = A_0) < \lim mg\ A = mg\ A_0).$$

9. If A is a formation of the first species,

$$(\lim A = A_0) \cap -(\omega A_0 = 0) < \left(\lim \frac{A}{\omega A} = \frac{A_0 \omega}{A_0} \right),$$

that is, if A has for its limit A_0, and the mass of A_0 is not zero, the simple point coincident with A has for its limit the simple point coincident with A_0.

10. $(\lim A = A_0) \cap (\lim a = a_0) \cap -(mg\ a_0 = 0) < \left(\lim \frac{mg\ Aa}{mg\ a} = \frac{mg\ A_0 a_0}{mg\ a_0} \right).$

Assuming A is a simple point, this proposition can be read: if the point A and the line a have A_0 and a_0, respectively, for limits, and the magnitude of a_0 is not zero, the distance of A from a has for its limit the distance of A_0 from a_0.

Thus, recalling the definitions of a right-line and of a plane (*Applicazione Geom.*, P. 30),[*] one deduces

11. $(\lim a = a_0) \cap -(mg\ a_0 = 0) <$ *the right-line a has for its limit the right-line* a_0.

And analogously

12. $(\lim \alpha = \alpha_0) \cap -(mg\ \alpha_0 = 0) <$ *the plane α has for its limit the plane α_0.*

66.   If $A(t)$ is a geometric formation of any species whatever, a function of a numerical variable t, we put

$$A'(t) = \lim \frac{1}{h}[A(t+h) - A(t)].$$

The formation $A'(t)$, which is of the same species as $A(t)$, is called the *derivative* of $A(t)$. We also indicate it with the expression $\frac{dA(t)}{dt}$; in this dt is an arbitrary number, called the differential of the independent variable; the formation $dA(t) =$

---

[*]G. Peano, *Applicazioni geometriche del Calcolo infinitesimale*, Torino, 1887.

$A'(t)dt$ is called the *differential* of $A(t)$. Instead of $A(t)$, substituting the variable t, we will often write just A.

One has the following rules for differentiation:

1. If A and B are variable formations of the same species, one has

$$d(A + B) = dA + dB.$$

2. $d(xA) = dx.A + x.dA.$

3. If AB is the progressive or regressive product of the two formations, one has

$$d.AB = dA.B + A.dB.$$

(one notes that the second member is not always equal to $BdA + AdB$).

Thus, if $ABC\ldots$ is the progressive or regressive product of the formations $ABC\ldots$, composed of products of two factors, one has

$$d(ABC\ldots) = (dA)BC\ldots + A(dB)C\ldots + AB(dC)\ldots + \ldots$$

4. If I is a vector in the plane, one has

$$d \perp I = \perp dI.$$

5. If I is a vector or bivector in space, one has

$$d|I = |dI.$$

6. If $x_1 x_2 \ldots$ are the coordinates of the variable formation A with respect to the fixed reference formations $A_1, A_2, \ldots$, that is if $A = x_1 A_1 + x_2 A_2 + \ldots$, one has $dA = dx_1 A_1 + dx_2 A_2 + \ldots$, that is the coordinates of the derivative of A are the derivatives of the coordinates of A.

Theorem. *If the derivative of $A(t)$ is zero, the function $A(t)$ is a constant.*

Indeed, if the derivative of A is zero, the coordinates of this derivative, that is the derivatives of the coordinates of A, are also zero; thus those coordinates are constants, and therefore A is constant.

67.   The derivative of the derivative of $A(t)$ is called the *second derivative* of $A(t)$ and is indicated by $A''(t)$, or $\frac{d^2A}{dt^2}$; the $(n-1)^{\text{th}}$ derivative of the derivative is called the nth derivative, and is indicated by $A^{(n)}(t)$ or by $\frac{d^nA}{dt^n}$. The formation $d^nA = A^{(n)}(t)dt^n$ is called the differential of order n.

One recognizes at once that the derivative of order n of a formation has for coordinates, with respect to fixed reference elements, the derivatives of order n of the coordinates of the formation given.

One also has

$$d^n.AB = d^n A.B + nd^{n-1}A.dB + \binom{n}{2} d^{n-2}A.d^2 B + \ldots + A.d^n B.$$

Theorem (of Taylor). *If $A(t)$ is a formation function of* t, *having the derivatives that appear in the following formula, determining R in a way such that one has*

$$A(t+h) = A(t) + hA'(t) + \frac{h^2}{2!}A''(t) + \ldots + \frac{h^n}{n!}A^{(n)}(t) + R,$$

one has, for $h = 0$,

$$\lim \frac{R}{h^n} = 0.$$

68.   Definition. We will say that the number a is *medial* {medio} between the numbers $a_1 a_2 \ldots$, finite or infinite in number, if it is neither greater than the limit superior of those numbers nor less than their limit inferior.

We will say that the volume A is *medial* between the volumes $A_1 A_2 \ldots$ if, Ω being the unit volume, the number $\frac{A}{\Omega}$ is medial between the numbers $\frac{A_1}{\Omega}, \frac{A_2}{\Omega}, \ldots$.

We will say that the formation A, of 1st, or 2nd, or 3rd species, is *medial* between the formations $A_1 A_2 \ldots$ of the same species as A, if, however one may take the surface, or line, or point, P, the volume AP is medial between the volumes $A_1 P, A_2 P, \ldots$.

For example, if $A_1 A_2$ are formations of the same species, $m_1, m_2, \ldots$ all being positive numbers, the formation $\frac{(m_1 A_1 + m_2 A_2 + \ldots)}{m_1 + m_2 + \ldots}$ is medial between the formations $A_1, A_2, \ldots$.

If $A_1 A_2 \ldots$ are simple points, the formations medial between them are the simple points belonging to the minimum closed convex field that contains them (Vide *Appl. Geom.*, P. 152 and 164).

If A is medial between $A_1 A_2 \ldots$, the coordinates of A are medial between those of $A_1, A_2, \ldots$, but not vice-versa.

Theorem. *If $A(t)$ has derivatives for all the values of* t *in the interval* (t_0, t_1), *then* $\frac{(A(t_1) - A(t_0))}{t_1 - t_0}$ *is medial between the values of $A'(t)$ in the same interval.*

Corollary. *If $A(t)$ has a derivative in the interval* (t_0, t_1), *and one sets it to zero at the extrema of this interval, a value medial between those of the derivative is zero.*

Theorem. *If $A(t)$ has successive derivatives up to order* $n + 1$ *in the interval* $(t, t + h)$, *the formation K that satisfies the condition*

$$A(t+h) = A(t) + hA'(t) + \ldots + \frac{h^n}{n!}A^{(n)}(t) + \frac{h^{n+1}}{(n+1)!}K$$

is medial between the values of the derivative of order $n + 1$ *in the same interval.*

Thus one has a second form of the theorem of Taylor.

69. Definition. We will put

$$A(t_1, t_2) = \frac{A(t_2) - A(t_1)}{t_2 - t_1},$$

$$A(t_1, t_2, t_3) = \frac{A(t_1, t_3) - A(t_1, t_2)}{t_3 - t_2}, \ldots.$$

The functions $A(t_1, t_2)$, $A(t_1, t_2, t_3)$, ... are called the interpolating functions of first, second, etc. order (Vide *Calc. diff.*, N. 81).*

Theorem 1. *The interpolating function of order* n, $A(t_0 t_1 \ldots t_n)$, *is medial between the values of* $\frac{A^{(n)}(t)}{n!}$, *where t is included among* $t_0, t_1, \ldots, t_n$.

Theorem 2. *If* $A^{(n)}(t)$ *is a continuous function in t, making* $t_0 t_1 \ldots t_n$ *tend to the same value t, one has*

$$\lim A(t_0, t_1, \ldots, t_n) = \frac{A^{(n)}}{n!}.$$

Theorem 3. *If* $A(t)$ *is a formation of the first species, having continuous successive derivatives, making* $t_1 t_2 t_3 \ldots$ *tend toward the same value t, one has*

(a)
$$\lim \frac{(A(t_1)A(t_2)}{t_2 - t_1} = A(t)A'(t)$$

(b)
$$\lim \frac{A(t_1)A(t_2)A(t_3)}{(t_2 - t_1)(t_3 - t_1)(t_3 - t_2)} = \frac{1}{2}A(t)A'(t)A''(t)$$

(c)
$$\lim \frac{A(t_1)A(t_2)A(t_3)A(t_4)}{(t_2 - t_1)(t_3 - t_1)(t_3 - t_2)(t_4 - t_1)(t_4 - t_2)(t_4 - t_3)}$$
$$= \frac{1}{2!3!}A(t)A'(t)A''(t)A'''(t).$$

In fact, to demonstrate (b), one has

$$A(t_1)A(t_2)A(t_3) = A(t_1)[A(t_2) - A(t_1)][A(t_3) - A(t_1)],$$

whence

$$\frac{A(t_1)A(t_2)A(t_3)}{(t_2 - t_1)(t_3 - t_1)} = A(t_1)A(t_1 t_2)A(t_1 t_3) = A(t_1)A(t_1 t_2)[A(t_1 t_3) - A(t_1 t_2)];$$

---

*G. Peano (A. Genocchi), *Calcolo differenziale e principii di calcolo integrale*, Torino, 1884.

thus, dividing by $t_3 - t_2$ as well,

$$\frac{A(t_1)A(t_2)A(t_3)}{(t_2 - t_1)(t_3 - t_1)(t_3 - t_2)} = A(t_1)A(t_1t_2)A(t_1t_2t_3).$$

Passing to the limit, by virtue of the preceding theorem, one has the formula to be demonstrated.

Theorem. *If $a(t)$ is a formation of the second species in the plane, one has*

$$\lim \frac{a(t_1)a(t_2)}{(t_2 - t_1)} = a(t)a'(t),$$

$$\lim \frac{a(t_1)a(t_2)a(t_3)}{(t_2 - t_1)(t_3 - t_1)(t_3 - t_2)} = \frac{1}{2}a(t)a'(t)a'''(t).$$

Theorem. *If α is a formation of the third species in space, one has*

$$\lim \frac{\alpha(t_1)\alpha(t_2)}{(t_2 - t_1)} = \alpha(t)\alpha'(t),$$

$$\lim \frac{\alpha(t_1)\alpha(t_2)\alpha(t_3)}{(t_2 - t_1)(t_3 - t_1)(t_3 - t_2)} = \frac{1}{2}\alpha(t)\alpha'(t)\alpha''(t),$$

$$\lim \frac{\alpha(t_1)\alpha(t_2)\alpha(t_3)\alpha(t_4)}{(t_2 - t_1)\dots(t_4 - t_3)} = \frac{1}{12}\alpha(t)\alpha'(t)\alpha''(t)\alpha'''(t).$$

70.   Definition. By the expression $\int A(t)dt$ we mean a function having $A(t)$ for derivative.

If $B(t)$ is a function having $A(t)$ for derivative, every other function having the same derivative is of the form $B(t) + C$, where C is a constant formation, of the same species as B. One demonstrates, upon reducing it to its correspondent on numerical functions, the proposition: If $A(t)$ is a continuous function in t, it is integrable.

The rules of integration that follow are recovered immediately from those seen for derivatives:

1. $\int (A + B)dt = \int A\,dt + \int B\,dt.$
2. $\int x\,dA = xA - \int A\,dx.$
3. $\int A\,dB) = AB - \int dA.B).$
4. $\int \perp I\,dt = \perp \int I\,dt.$
5. $\int |I\,dt = |\int I\,dt.$
6. *If $A_1A_2\dots$ are constant functions, and $x_1, x_2, \dots$ numerical functions of t, one has*

$$\int (x_1A_1 + x_2A_2 + \dots)dt = A_1 \int x_1\,dt + A_2 \int x_2\,dt + \dots.$$

Thus the coordinates, with respect to fixed elements, of the integral of a geometric formation are the integrals of the coordinates of the formation proposed.

Definition. If $B(t)$ is one of the integrals of $A(t)$, we put

$$\int_{t_0}^{t_1} A(t)dt = B(t_1) - B(t_0).$$

Theorem. *If $A(t)$ has successive derivatives, one has*

$$A(t+h) = A(t) + hA'(t) + \frac{h^2}{2!}A''(t) + \ldots + \frac{h^n}{n!}A^{(n)}(t)$$

$$+ \frac{h^{n+1}}{n!}\int_0^1 (1-z)^n A^{(n+1)}(t+zh)dz).$$

This theorem is a third form of the theorem of Taylor applied to geometric formations.

Applications

71.   1. Let $A(t)$ be a formation of first species function of the variable t, having continuous successive derivatives, and suppose that, for those values of t under consideration, the mass of A is not zero. Then the simple point coincident with A, upon varying t, describes a line. The right-line $A(t_1)A(t_2)$ connects the two points of the curve corresponding to the values t_1 and t_2 of t. Making t_1 and t_2 tend to the same value t, by virtue of Theorem 3, of N. 68 and 11 of N. 64, one deduces that the preceding right-line has for limit the right-line $A(t)A'(t)$, assuming AA' is not zero. This limiting right-line is the *tangent* to the curve at the point considered.

2. The plane $A(t_1)A(t_2)A(t_3)$ contains three points of the curve; making $t_1t_2t_3$ tend to the same value t, it has for limit the plane $A(t)A'(t)A''(t)$, assuming that that formation is not zero. This limiting plane is the *osculating plane* of the curve.

3. If $A(t)$ is a simple point, its successive derivatives are vectors. If the variable t represents the *time*, the first derivative is called the *velocity*, the second the *acceleration* of the point. If $A(t)$ is a point with constant mass, its derivatives are also vectors; if t is the time, the first derivative is called the *quantity of motion* of the material point, the second the *force* that acts on the point.

4. If $A(t)$ is a simple point moved in a fixed plane, the right-line $A \perp A'$ is the normal to the curve described by A.

5. If $A(t)$ is a simple point that is moved in space, then the direction of the vector A' is that of the *tangent* AA'; and the position of the bivector $A'A''$ is that of the *osculating* plane $AA'A''$. The position of $|A'$ is normal to the direction of the tangent; the plane $A|A'$ is the *normal plane* to the curve. The direction of $|(A'A'')$ is normal to the osculating plane; the right-line $A|(A'A'')$ is the *binormal*. The direction of $A'A''.|A' = (A'|A').A'' - (A'|A'').A'$ is that of the intersection of the positions of the osculating and normal planes; the right-line $A(A'A''.|A') = (A'|A').AA'' - (A'|A'')AA'$ is the *principal normal*.

6. The integral $\int (mg\ A)dt = \int (mg\ dA)$, taken between definite limits, represents the *arc length* described by the simple point A, where t varies between the fixed limits. Calling s the arc length described by A, where t varies from a fixed value to a variable value t, one has $ds = (mg\ A)dt = mg\ dA$.

7. The integral $\int AA'dt = \int A\,dA$ represents a formation of second species. It is the limit toward which tends the sum $A_0A_1 + A_1A_2 + \dots + A_{n-1}A_n$, in which $A_0A_1 \dots A_n$ are consecutive points on the curve, A_0 and A_n are the positions of A corresponding to the limit of t, the limit being attained upon making the number of points increase indefinitely, in such a way that the distance between any two consecutive points tends to zero.

The vector of $\int AA'dt$ is $\int A'\,dt$, that is, it is the vector formed by the extrema of the arc considered.

8. If the point A is moved in a plane and describes an arc whose extrema do not coincide, the formation $\int AA'\,dt$ is reducible to a line MN; if X is any point in the plane, the area described by the segment XA, where A describes the arc of the curve considered, is equal to the area of the triangle XMN. If however the extrema of the arc considered coincide, $\int AA'\,dt$ is a bivector, and the area $X\int AA'\,dt$, which is independent of the position of the point X, is called the enclosed area of that closed curve.

9. If the point A is moved in space, and the extrema of the arc considered coincide, the formation $\int AA'\,dt$ is reducible to a bivector i. If XY are two points in space, the volume described by the triangle XYA, where A describes the closed arc considered, is equal to the volume XYi. If one projects on a plane, with parallel projections, the closed line described by A and the bivector i, the enclosed area of the projection of that line has for projection the projection of i.

10. The integral $\int A(mg\ A')dt = \int A(mg\ dA)) = \int A\,ds$ represents a formation of first species, with mass equal to the arc length described by A. It is the limit toward which tends the formation $s_1A_1 + s_2A_2 + \dots + s_nA_n$, where $s_1s_2 \dots s_n$ are the lengths of the n arcs into which the given arc is decomposed, and $A_1A_2 \dots A_n$ are points on these partial arcs. The simple point that coincides with this formation, that is $\frac{\int A(mg\ dA)}{\int (mg\ dA)}$, is called the *barycenter* of the arc.

11. The point $A = O + tI + \frac{1}{t}J$, where A is a simple point, I and J are vectors, describes an *hyperbola* with center O and asymptotes OI and OJ. One has: $A' = I - \frac{1}{t^2}J$; thus one perceives that the four vectors $I, J, A - O, A'$ are harmonic. One has the tangent $AA' = OI - \frac{1}{t^2}OJ - \frac{1}{t}IJ$. Calling B and C the points of intersection with the asymptotes OI and OJ, one has $B = \frac{AA'.OI}{AA'.OI.u} = O + 2tI$; and analogously $C = O + \frac{2}{t}J$.

One deduces immediately the formulas $A = \frac{1}{2}(B+C)$, $OBC = 4OIJ$, which, interpreted geometrically, express known properties of the hyperbola.

The area described by the segment OA, while t varies from t_0 to t_1 is given by the integral $\int_{t_0}^{t_1} OAA' \, dt = \int_{t_0}^{t_1} -\frac{2}{t}OIJ \, dt = 2OIJ \log \frac{t_0}{t_1}$.

12. The point $O + It + Jt^2$, where O is a simple point, I and J are vectors, describes a parabola.

The point $O + I \cos t + J \sin t$, as also $O + I \frac{1-t^2}{1+t^2} + J \frac{2t}{1+t^2}$, where OIJ have the same meaning as in the preceding examples, describe ellipses. One can apply to these lines the formulas found.

13. The point $O + r \cos t.I + r \sin t.J + ht.K$, where O is a simple point, and IJK are three vectors equal to the unit of measure and orthogonal pairwise, describe a helix having the right-line OK as axis, radius r, pitch $2\pi h$. The closed line l formed of one coil of the helix, the two radii of the cylinder that go to the extrema of the coil, and the portion of the axis between those two radii, considered as a formation of second species, is equal to the bivector of the circular base of the cylinder on which the helix is traced. Thus, projecting with parallel right-lines on an arbitrary plane, the enclosed area of the projection of the line l is equal to the area of the projection of the circular base.

14. Let $a(t)$ be a line function of the variable t, which is moved in a fixed plane. The limit of the point $a(t_1)a(t_2)$ of intersection of the two lines a, where t_1 and t_2 tend to the same value t, is the point aa', supposing aa' is not zero. This point is called the point of intersection of two consecutive lines of the system, or the point of contact of a with its envelope. The right-line $aa'. \perp ua$ is the *normal to the envelope*.

15. Let $A(t)$ be a simple point, mobile in the plane. Setting $a = A \perp A'$, the right-line a is the normal to the curve described by A. One has $a' = A' \perp A' + A \perp A''$. Thus the point aa' of intersection of two consecutive normals can be reduced, after some transformations, to the form $A + \frac{mg^2 A'}{A'A''}. \perp A'$.

16. Let $a(t)$ be a line that is moved in space, and having successive derivatives, which are formations of second species. Since a is a line, one has $aa = 0$, whence, differentiating, $aa' = 0$, $aa'' + a'a' = 0$, etc. Upon varying t, the line a describes a *ruled surface*. If P is a point of the right-line a, the plane Pa' is the plane tangent to the surface at the point P; or if π is a plane passing through the right-line a, the point $\pi a'$ is the point of contact of the plane π with the surface.

Giving to t the values t_1 and t_2, calling a_1 and a_2 the corresponding positions of a, θ the angle between them, δ their minimum distance, one recovers

$$\lim \frac{\theta}{t_2 - t_1} = \frac{mg(\omega a . \omega a')}{mg^2 a}, \quad \lim \frac{\delta}{t_2 - t_1} = \frac{a'a'}{2mg(\omega a . \omega a')}.$$

The limit of the plane passing through a_1 and parallel to a_2 is the plane $p.p'\omega = -p'.p\omega$. The limit of the extrema of the minimum distance between the two right-lines a_1 and a_2 is the point $[p|(\omega p.\omega p')]p'$.

17. Let $\alpha(t)$ be a formation of the third species function of t. If the bivector of α is not zero, this formation is reducible to a triangular surface. The right-line $\alpha\alpha'$, assuming $\alpha\alpha'\omega$ is not zero, is the limiting position of the intersection of the two planes α; the point $\alpha\alpha'\alpha''$, assuming $\alpha\alpha'\alpha''\omega$ is not zero, is the limit of the point of intersection of the three planes of the system, corresponding to values of t that tend to the same value. If $\alpha\alpha'\alpha''\alpha'''$ is not zero, the point $\alpha\alpha'\alpha''$ describes a curve whose tangent is $\alpha\alpha'$ and whose osculating plane is α. The normal plane is $\alpha\alpha'\alpha''.|(\alpha\alpha'\omega)$.

18. A geometric formation A may be a function of two or more numbers u, v,...; and then one has to consider the partial derivatives $\frac{dA}{du}$, $\frac{dA}{dv}$, $\frac{dA^2}{du^2}$, ..., for which there exist theorems analogous to those demonstrated for numerical functions of several variables.

19. If the formation of the first species A is a function of two variables u and v, and its mass is not zero, the simple point coincident with it, upon variation of u and v, describes a surface. The plane $A \frac{dA}{du} \frac{dA}{dv}$ is the *tangent plane* to the surface at the point A.

20. If A is a simple point function of u and v, its derivatives are vectors; the position $\frac{dA}{du} \frac{dA}{dv}$ is that of the tangent plane; the right-line $A| \left(\frac{dA}{du} \frac{dA}{dv} \right)$ is the normal to the surface.

21. Varying u between the fixed values u_0 and $u_1 > u_0$, and v between the values v_0 and $v_1 > v_0$, functions of u, the simple point A, a function of u and v, describes a bounded surface σ; by $\int \psi \, du \, dv$, where ψ is a function of u and v, we mean $\int_{u_0}^{u_1} \cdot \left(\int_{v_0}^{v_1} \psi \, dv \right)$. Then $2 \int \int \frac{dA}{du} \frac{dA}{dv} \, du \, dv$ represents a bivector equal to the contour {contorno} of σ. If one decomposes the surface σ into parts $\sigma_1 \sigma_2 \ldots \sigma_n$, calling $i_1 i_2 \ldots i_n$ their bivectors, the limit superior of the quantity $mg \, i_1 + mg \, i_2 + \ldots + mg \, i_n$ is called the *area* of the given surface. The area is measured by $2 \int \int mg \left(\frac{dA}{du} \frac{dA}{dv} \right) \, du \, dv$ (the unit of measure of area being that defined in N. 39).

22. $2 \int \int A \frac{dA}{du} \frac{dA}{dv} dudv$ represents a formation of the third species; it is the limit toward which tends the sum $A_1 i_1 + A_2 i_2 + \ldots + A_n i_n$, where $i_1 i_2 \ldots i_n$ are the bivectors of the parts $\sigma_1 \sigma_2 \ldots \sigma_n$ into which one can decompose the surface σ, $A_1 A_2 \ldots A_n$ are points of these parts, the limit being attained by varying the decomposition of σ in such a way that the distance between two points belonging to the same part tends to zero. If the bivector of this formation is not zero, it is reducible to a triangular surface PQR. However one may take the point X in space, the volume described by the segment XA, where the point A describes the surface σ, taking account of the sign, is equal to the tetrahedron $XPQR$. If however that bivector is zero, the volume described by XA is independent of the point X.

23. $2 \int \int A mg \left(\frac{dA}{du} \frac{dA}{dv} \right) dudv$ represents a formation of the first species, whose mass is the area of the surface; the simple point coincident with it is called the *barycenter* of the surface.

24. If A is a simple point function of the three numbers t, u, v, then $6 \int \int \frac{dA}{dt} \cdot \frac{dA}{du} \cdot \frac{dA}{dv} dtdudv$ measures the volume described by the point A while the variables vary between the limits to which the integration extends.

If a rectilinear segment AB varies with the variation of the two numbers u and v, calling I the vector $B - A$ of that segment, and C its medial point, the volume described by AB is measured by $6 \int \int I \frac{dC}{du} \frac{dC}{dv} dudv + \frac{1}{2} \int \int I \frac{dI}{du} \frac{dI}{dv} dudv$.

If a plane area σ is moved with the variation of a number t, calling i the bivector of that area, and G its barycenter, the volume described by σ is measured by $3 \int idG$.

CHAPTER IX

Transformations
of Linear Systems

72. Let there exist a system of entities for which are given the following definitions:

1. The *equivalence* of two entities **a** and **b** of the system is defined, that is, a proposition, indicated by **a** = **b**, is defined, which expresses a condition between two entities of the system, satisfied by certain pairs of entities, and not by others, and which satisfies the logical equations:

$$(\mathbf{a} = \mathbf{b}) = (\mathbf{b} = \mathbf{a}), \quad (\mathbf{a} = \mathbf{b}) \cap (\mathbf{b} = \mathbf{c}) < (\mathbf{a} = \mathbf{c}).$$

2. The *sum* of two entities **a** and **b** is defined, that is to say an entity, indicated by **a** + **b**, is defined that also belongs to the system given, and which satisfies the conditions

$$(\mathbf{a} = \mathbf{b}) < (\mathbf{a} + \mathbf{c} = \mathbf{b} + \mathbf{c}), \quad \mathbf{a} + \mathbf{b} = \mathbf{b} + \mathbf{a}, \quad \mathbf{a} + (\mathbf{b} + \mathbf{c}) = (\mathbf{a} + \mathbf{b}) + \mathbf{c},$$

the common value of the two members of the last equivalence being indicated by **a** + **b** + **c**.

3. If **a** is an entity of the system, and m is a positive integer, by the expression m**a** we will mean the sum of m entities equal to **a**. It is easy to recognize that, if **a**, **b**, ... are entities of the system, and m, n,... positive integers,

$$(\mathbf{a} = \mathbf{b}) < (m\mathbf{a} = m\mathbf{b}); \quad m(\mathbf{a} + \mathbf{b}) = m\mathbf{a} + m\mathbf{b}; \quad (m + n)\mathbf{a} = m\mathbf{a} + n\mathbf{a};$$

$$m(n\mathbf{a}) = (mn)\mathbf{a}; \quad 1\mathbf{a} = \mathbf{a}.$$

We will suppose that a meaning is attributed to the expression m**a**, whatever may be the real number m, in such a way that the preceding equations are also satisfied. The entity m**a** is called the *product* of the (real) number m with the entity **a**.

4. Finally we will suppose there exists an entity of the system, which we will call the *null entity*, and that we will indicate by 0, such that whatever may be the entity **a**, the product of the number 0 with the entity **a** always gives the entity 0, that is

$$0\mathbf{a} = 0.$$

If to the expression $\mathbf{a} - \mathbf{b}$ one attributes the meaning $\mathbf{a} + (-1)\mathbf{b}$, one deduces

$$\mathbf{a} - \mathbf{a} = 0, \qquad \mathbf{a} + 0 = \mathbf{a}.$$

Definition. The system of entities for which definitions 1, 2, 3, 4 are given, in such a way as to satisfy the conditions imposed, is called a *linear system*.

One deduces that if $\mathbf{a}$, $\mathbf{b}$, $\mathbf{c}$, ... are entities of the same linear system, m, n, p, ... real numbers, every linear homogeneous function of the form $m\mathbf{a} + n\mathbf{b} + p\mathbf{c} + \ldots$ represents an entity of the same system.

Real numbers, and formations of the same species in space, constitute linear systems.

In addition, formations of the first species on a right-line, or in the plane, vectors in a plane or in space, and so on, also constitute linear systems. But the points of space do not constitute a linear system, since their sums, following the definitions given, are not more points, but any formation of the first species.

73.   Definition. Several entities $\mathbf{a}_1 \mathbf{a}_2 \ldots \mathbf{a}_n$ of a linear system are called mutually *dependent* if one can determine n numbers $m_1 m_2 \ldots m_n$, not all zero, for which there results

$$m_1\mathbf{a}_1 + m_2\mathbf{a}_2 + \ldots + m_n\mathbf{a}_n = 0.$$

In this case any one of the entities whose coefficient is not zero can be expressed as a linear homogeneous function of the rest.

If the entities $\mathbf{a}_1 \ldots \mathbf{a}_n$ are mutually independent, and if between them there is a relation $m_1\mathbf{a}_1 + \ldots m_n\mathbf{a}_n = 0$, one deduces $m_1 = 0, \ldots m_n = 0$.

If $AB \ldots$ are formations of first species in space, the equations $AB = 0$, $ABC = 0$, $ABCD = 0$ express the dependence of 2 or 3 or 4 formations. 5 formations of the first species in space are always mutually dependent.

Definition. The number of dimensions of a linear system is the maximum number of mutually independent entities of the system that one can take.

For example, the formations of the first species on a right-line, or in the plane, or in space, form linear systems of 2, 3, and 4 dimensions, respectively; the vectors in the plane or in space form systems of 2 and 3 dimensions; the formations of the second species in space form a system of 6 dimensions. The real numbers form a linear system of one dimension; the imaginary numbers, or ordinary complexes, form a system of two dimensions. A linear system can also have infinite dimensions.

Theorem. *If the system A is of* n *dimensions, taking* n *independent entities* $\mathbf{a}_1 \ldots \mathbf{a}_n$ *of the system, and given a new entity* $\mathbf{a}$, *one can always determine* n *numbers* $x_1 \ldots x_n$ *for which there results*

(1) $$\mathbf{a} = x_1\mathbf{a}_1 + \ldots + x_n\mathbf{a}_n.$$

In addition they are determined uniquely, that is

(2) $$(x_1\mathbf{a}_1 + \ldots + x_n\mathbf{a}_n = x_1'\mathbf{a}_1 + \ldots + x_n'\mathbf{a}_n) = (x_1 = x_1') \cap \ldots \cap (x_n = x_n').$$

In fact, since the system A is of n dimensions, between the $n + 1$ entities $\mathbf{a}, \mathbf{a}_1, \ldots , \mathbf{a}_n$ there is a relation of the form of definition 1; in this relation the coefficient of $\mathbf{a}$ is not zero, for otherwise there would be a relation between $\mathbf{a}_1, \ldots , \mathbf{a}_m$, contrary to hypothesis; thus from that relation with respect to $\mathbf{a}$ one has the formula to be demonstrated.

If one puts $x_1\mathbf{a}_1 + \ldots = x_1'\mathbf{a}_1 + \ldots$, one deduces $(x_1 - x_1')\mathbf{a}_1 + \ldots = 0$; and thus, since $\mathbf{a}_1, \ldots$ are independent, $x_1 = x_1', x_2 = x_2' \ldots$.

Definition. If $\mathbf{a}_1 \ldots \mathbf{a}_n$ are independent entities of a system of n dimensions, the numbers $x_1 \ldots x_n$ that satisfy relations (1) are called the *coordinates* of $\mathbf{a}$ with respect to the *reference entities* $\mathbf{a}_1 \ldots \mathbf{a}_n$.

The formulas

(3) $$(x_1\mathbf{a}_1 + \ldots + x_n\mathbf{a}_n) + (y_1\mathbf{a}_1 + \ldots + y_n\mathbf{a}_n)$$
$$= (x_1 + y_1)\mathbf{a}_1 + \ldots + (x_n + y_n)\mathbf{a}_n,$$

(4) $$m(x_1\mathbf{a}_1 + \ldots + x_n\mathbf{a}_n) = mx_1\mathbf{a}_1 + \ldots + mx_n\mathbf{a}_n,$$

give the coordinates of the sum of two entities, and the product of a number m with an entity as functions of the coordinates of that entity. The coordinates of the entity 0 are all zero.

The preceding definitions coincide with the definitions given in N. 34, 38, 46, 50, 60 for formations of the first species on a right-line, vectors in the plane, formations of the first and second species in the plane, etc.

Assume $x_1 \ldots x_n$ are the coordinates of $\mathbf{a}$ with respect to the reference elements $\mathbf{a}_1 \ldots \mathbf{a}_n$, that is

$$\mathbf{a} = x_1\mathbf{a}_1 + \ldots + x_n\mathbf{a}_n,$$

and one knows the coordinates of the $\mathbf{a}_1 \ldots \mathbf{a}_n$ with respect to another group of reference entities $\mathbf{b}_1 \ldots \mathbf{b}_n$, that is

$$\mathbf{a}_1 = m_{11}\mathbf{b}_1 + \ldots + m_{1n}\mathbf{b}_n, \ldots , \mathbf{a}_n = m_{n1}\mathbf{b}_1 + \ldots + m_{nn}\mathbf{b}_n;$$

then upon substituting one deduces

$$\mathbf{a} = (m_{11}x_1 + \ldots + m_{n1})\mathbf{b}_1 + \ldots + (m_{1n}x_1 + \ldots + m_{nn}x_n)\mathbf{b}_n.$$

One has in this way calculated the coordinates of $\mathbf{a}$ with respect to $\mathbf{b}_1 \ldots \mathbf{b}_n$; they are linear functions and homogeneous in the coordinates of $\mathbf{a}$ with respect to $\mathbf{a}_1 \ldots \mathbf{a}_n$; the coefficients of those functions are the coordinates of $\mathbf{a}_1 \ldots \mathbf{a}_n$ with respect to $\mathbf{b}_1 \ldots \mathbf{b}_n$.

74.   An entity of a system may be considered as a constant or a variable.

Definition. We will say that a variable entity $\mathbf{a}$ of a linear system, of a finite number of dimensions, has for limit the fixed entity $\mathbf{a}_0$, if the coordinates of $\mathbf{a}$ with respect to fixed entities, have for limits the coordinates of $\mathbf{a}_0$.

It follows from the previous results that the limit does not depend on the reference elements.

A variable entity $\mathbf{a}$ of a linear system may be a function of a numerical variable t; for that function one can give the definitions of the *derivative, successive derivative, definite and indefinite integral* that one recovered in N. 66, 67 and 70, reading 'entity of a linear system' instead of 'geometric formation'; also applicable are the theorems found there that refer only to sums, multiplication by a number, and coordinates.

An entity $\mathbf{y}$, a function of an entity $\mathbf{x}$, is called a *continuous function*, if the limit of the function is the function of the limit. Naturally one supposes defined the limit of the entity $\mathbf{x}$ and the entity $\mathbf{y}$. If the entity $\mathbf{y}$ of a linear system of a finite number of dimensions is a continuous function of the entity $\mathbf{x}$ belonging only to a linear system of a finite number of dimensions, the coordinates of $\mathbf{y}$ are continuous (numerical) functions of the coordinates of $\mathbf{x}$.

75.   Definition. An operation $\mathbf{R}$, to be carried out on every entity $\mathbf{a}$ of a linear system A, is called *distributive*, if the result of the operation $\mathbf{R}$ on the entity $\mathbf{a}$, which we will indicate by $\mathbf{Ra}$, is also an entity of a linear system, and the identities

$$\mathbf{R}(\mathbf{a} + \mathbf{a}') = \mathbf{Ra} + \mathbf{Ra}', \qquad \mathbf{R}(m\mathbf{a}) = m(\mathbf{Ra})$$

are verified, where $\mathbf{a}$ and $\mathbf{a}'$ are any entities whatever of the system A, and m is any real number.

The entity $\mathbf{Ra}$, that is the result of the distributive operation $\mathbf{R}$ on the entity $\mathbf{a}$, is called a *distributive function* of $\mathbf{a}$. A distributive operation is also called a *linear transformation*, or *transformation* without any qualifier. If the entity $\mathbf{Ra}$ belongs to the same linear system A as does $\mathbf{a}$, the operation $\mathbf{R}$ is called a *substitution*. The entity $\mathbf{Ra}$ is also called the product of the transformation $\mathbf{R}$ on the entity $\mathbf{a}$.

If **R** is a distributive operation, **a**, **b**, ... entities of the system A, and m, n, ... are numbers, one has

$$\mathbf{R}(m\mathbf{a} + n\mathbf{b} + \dots) = m\mathbf{Ra} + n\mathbf{Rb} + \dots$$

$$\mathbf{R}0 = 0.$$

The unique distributive operation to be carried out on (real) numbers, and whose result is again a number, is multiplication. If A and B are geometric formations of any species, their progressive or regressive product AB is a distributive function of both factors. The operations indicated by the symbols $\perp$ and $|$, which are to be carried out on vectors in the plane, or on vectors and bivectors in space, are distributive operations.

Multiplication of an entity of a linear system by a number m is a distributive operation. Any one of the coordinates of the entity **a** of a system of n dimensions, with respect to fixed entities, is a distributive function of **a**, as indicated by formulas (3) and (4) of N. 73.

One observes that, of the two conditions imposed on an operation in order that it be distributive, the second is a consequence of the first for commensurable m; if m is incommensurable, it can also be deduced from the first, assuming, e.g., the continuity of the formation **Ra**.

76.   We will introduce the following conventions:

1. If **R** and **S** are two transformations of the entities of the system A into entities of the same linear system B, we will put

$$\mathbf{R} = \mathbf{S}$$

if, whatever may be the entity **a** of the system A, one has **Ra** = **Sa**.

2. The multiplication of the entities of a system by a number m, is, as has been said, a distributive operation; we indicate it by the same number m. Thus the equivalences **R** = 1, **R** = 0 say that the operation **R** makes correspond to each entity itself, or zero, respectively.

3. If **R** and **S** are transformations of the entities **a** of a system A into entities of the same linear system B, we will put

$$(\mathbf{R} + \mathbf{S})\mathbf{a} = \mathbf{Ra} + \mathbf{Sa}.$$

The operation indicated by the symbol **R** + **S** is also distributive. We will call it the *sum* of the two operations **R** and **S**; it is a new transformation of the entities of A into entities of B. It evidently satisfies the conditions imposed on a sum in N. 72.

4. If $\mathbf{R}$ is a transformation of the entities $\mathbf{a}$ of the system A into entities of the system B, and if $\mathbf{S}$ is a transformation of the entities of B into entities of a system C, we will put

$$\mathbf{SRa} = \mathbf{S(Ra)}.$$

The operation indicated by the symbol $\mathbf{SR}$ is also distributive; we will call it the *product* of the two transformations $\mathbf{R}$ and $\mathbf{S}$ is defined. The operation $\mathbf{SR}$ transforms the entities of the system A into entities of the system C.

As a particular case, since a number represents a transformation, the product $m\mathbf{R}$, of a transformation $\mathbf{R}$ with a number is defined, and it satisfies the conditions imposed in N. 72, 3.

One deduces that the various transformations of the entities of a system A into entities of a system B constitute a linear system.

If $\mathbf{R}, \mathbf{R}'$ are transformations of the entities A into entities B, and if $\mathbf{S}, \mathbf{S}'$ are tranformations of the entities B into entities C, one has

$$\mathbf{S(R + R')} = \mathbf{SR} + \mathbf{SR'}, \qquad \mathbf{(S + S')R} = \mathbf{SR} + \mathbf{S'R}.$$

If $\mathbf{R}, \mathbf{S}, \mathbf{T}$, respectively, transform the entities A into B, the entities B into C, and the entities C into D, one has $\mathbf{T(SR)} = \mathbf{(TS)R}$, and their common value is indicated by $\mathbf{TSR}$; this is a transformation of the entities of the system A into entities of the system D.

Thus the multiplication of transformations has the distributive property with respect to the sum, and associativity; but it is not in general commutative, that is, one does not in general have $\mathbf{SR} = \mathbf{RS}$. When this condition *is* satisfied, the two operations are called mutually *commutative*. If m is a number, $\mathbf{R}$ any transformation, one has $m\mathbf{R} = \mathbf{R}m$, that is the transformations equal to a number are commutative with every other transformation.

77.   Theorem. *If $\mathbf{a}_1 \ldots \mathbf{a}_n$ are independent entities of a linear system A of* n *dimensions, and $\mathbf{b}_1 \ldots \mathbf{b}_n$ are also entities of a linear system, one and only one transformation $\mathbf{R}$ of the system A is determined that satisfies the conditions*

(α) $\qquad\qquad\qquad \mathbf{Ra}_1 = \mathbf{b}_1, \ldots , \mathbf{Ra}_n = \mathbf{b}_n.$

In fact, to each entity $\mathbf{a}$ of the system A corresponds n numbers $x_1 \ldots x_n$ that satisfy the condition $\mathbf{a} = x_1\mathbf{a}_1 + \ldots + x_n\mathbf{a}_n$, that is the coordinates of $\mathbf{a}$ with respect to $\mathbf{a}_1 \ldots \mathbf{a}_n$, which are distributive functions of $\mathbf{a}$. One puts $\mathbf{Ra} = x_1\mathbf{b}_1 + \ldots + x_n\mathbf{b}_n$; $\mathbf{Ra}$ is then a distributive function of $\mathbf{a}$, and thus the operation $\mathbf{R}$ is distributive, and evidently satisfies the condition (α) imposed. If then $\mathbf{R}$ and $\mathbf{S}$ are two transformations that satisfy condition (α), upon taking any entity $\mathbf{a} = x_1\mathbf{a}_1 + \ldots + x_n\mathbf{a}_n$ of A, one perceives at once that $\mathbf{Ra} = \mathbf{Sa}$, and thus $\mathbf{R} = \mathbf{S}$.

Definition. The transformation of the entities of the linear system A of n dimensions that makes correspond to the n independent entities $\mathbf{a}_1 \ldots \mathbf{a}_n$ the entities $\mathbf{b}_1 \ldots \mathbf{b}_n$ is indicated by the expression

$$\begin{pmatrix} \mathbf{b}_1 & \mathbf{b}_2 & \cdots & \mathbf{b}_n \\ \mathbf{a}_1 & \mathbf{a}_2 & \cdots & \mathbf{a}_n \end{pmatrix}.$$

The group of entities $\mathbf{a}_1 \ldots \mathbf{a}_n$ is also called the *denominator* of the transformation, while the group $\mathbf{b}_1 \ldots \mathbf{b}_n$ is called its *numerator*. For the denominator of a transformation one can take any group of n independent entities of A. Two transformations of the entities of the same system A can always be reduced to the same denominator.

One recovers from this definition and the preceding propositions:

$$\begin{pmatrix} \mathbf{b}_1 & \mathbf{b}_2 & \cdots & \mathbf{b}_n \\ \mathbf{a}_1 & \mathbf{a}_2 & \cdots & \mathbf{a}_n \end{pmatrix} (x_1 \mathbf{a}_1 + \ldots + x_n \mathbf{a}_n) = x_1 \mathbf{b}_1 + \ldots + x_n \mathbf{b}_n.$$

$$\left\{ \begin{pmatrix} \mathbf{b}_1 & \mathbf{b}_2 & \cdots & \mathbf{b}_n \\ \mathbf{a}_1 & \mathbf{a}_2 & \cdots & \mathbf{a}_n \end{pmatrix} = \begin{pmatrix} \mathbf{b}_1' & \mathbf{b}_2' & \cdots & \mathbf{b}_n' \\ \mathbf{a}_1 & \mathbf{a}_2 & \cdots & \mathbf{a}_n \end{pmatrix} \right\}$$

$$= (\mathbf{b}_1 = \mathbf{b}_1') \cap \ldots \cap (\mathbf{b}_n = \mathbf{b}_n').$$

$$\begin{pmatrix} \mathbf{b}_1 & \mathbf{b}_2 & \cdots & \mathbf{b}_n \\ \mathbf{a}_1 & \mathbf{a}_2 & \cdots & \mathbf{a}_n \end{pmatrix} + \begin{pmatrix} \mathbf{c}_1 & \mathbf{c}_2 & \cdots & \mathbf{c}_n \\ \mathbf{a}_1 & \mathbf{a}_2 & \cdots & \mathbf{a}_n \end{pmatrix}$$

$$= \begin{pmatrix} \mathbf{b}_1 + \mathbf{c}_1 & \mathbf{b}_2 + \mathbf{c}_2 & \cdots & \mathbf{b}_n + \mathbf{c}_n \\ \mathbf{a}_1 & \mathbf{a}_2 & \cdots & \mathbf{a}_n \end{pmatrix}.$$

$$\begin{pmatrix} m\mathbf{a}_1 & m\mathbf{a}_2 & \cdots & m\mathbf{a}_n \\ \mathbf{a}_1 & \mathbf{a}_2 & \cdots & \mathbf{a}_n \end{pmatrix} = m; \quad \begin{pmatrix} \mathbf{a}_1 & \cdots & \mathbf{a}_n \\ \mathbf{a}_1 & \cdots & \mathbf{a}_n \end{pmatrix} = 1;$$

$$\begin{pmatrix} 0 & 0 & \cdots & 0 \\ \mathbf{a}_1 & \mathbf{a}_2 & \cdots & \mathbf{a}_n \end{pmatrix} = 0.$$

The coordinate x_1 of the entity $\mathbf{a}$ with respect to the reference entities $\mathbf{a}_1 \ldots \mathbf{a}_n$ is, as has been said, a distributive function of $\mathbf{a}$; the operation that gives this coordinate can be indicated by the symbol

$$\begin{pmatrix} \mathbf{a}_1 & 0 & \cdots & 0 \\ \mathbf{a}_1 & \mathbf{a}_2 & \cdots & \mathbf{a}_n \end{pmatrix}.$$

If a, b, c are numbers, the expression $\begin{pmatrix} c \\ b \end{pmatrix}$ a equals $\frac{c}{b} a$.

Definition. If $\mathbf{R} = \begin{pmatrix} \mathbf{b}_1 & \cdots & \mathbf{b}_n \\ \mathbf{a}_1 & \cdots & \mathbf{a}_n \end{pmatrix}$ is a transformation of the entities of the system A of n dimensions into the entities of the system B, also of n dimensions, and the entities $\mathbf{b}_1 \ldots \mathbf{b}_n$ are mutually independent, we will put

$$\mathbf{R}^{-1} = \begin{pmatrix} \mathbf{a}_1 & \cdots & \mathbf{a}_n \\ \mathbf{b}_1 & \cdots & \mathbf{b}_n \end{pmatrix}.$$

$\mathbf{R}^{-1}$ is a transformation of the entities of the system B into entities of the system A; it is called the *inverse* of $\mathbf{R}$. One has

$$\mathbf{R}^{-1}\mathbf{R} = 1, \quad \mathbf{R}\mathbf{R}^{-1} = 1, \quad (\mathbf{R}^{-1})^{-1} = \mathbf{R}.$$

78.   Deserving special mention among the transformations are the *substitutions*, that is the transformations of the entities of a linear system A into entities of the same linear system. Transformations equal to numbers are substitutions. Substitutions can always be summed and multiplied among themselves and with numbers.

If $\mathbf{R}$ is a substitution, we will indicate by $\mathbf{R}^n$, where n is a positive integer, the product of n substitutions $\mathbf{R}$. One has

(1) $\mathbf{R}^m\mathbf{R}^n = \mathbf{R}^{m+n}; \quad (\mathbf{R}^m)^n = \mathbf{R}^{mn}.$

If the substitution $\mathbf{R}^{-1}$ inverse to $\mathbf{R}$ exists, and n is a positive integer, we will put $\mathbf{R}^{-n} = (\mathbf{R}^{-1})^n$, and $\mathbf{R}^0 = 1$. One deduces that, whatever the integers m and n may be, positive or negative, formula (1) holds.

If $\mathbf{R}$ is any substitution, we will put by definition

$$e^{\mathbf{R}} = 1 + \mathbf{R} + \frac{\mathbf{R}^2}{2!} + \frac{\mathbf{R}^3}{3!} + \cdots .$$

One supposes that there exist for linear entities and for transformations definitions of series, convergence, and their sums.

One can demonstrate that the series of the second member is always convergent, assuming that the linear system A has a finite number of dimensions (Vide {G. Peano}, Integrazione per serie delle equazioni differenziali lineari. *Atti dell'Acc. delle Scienze di Torino*, 1887).

If t is a numerical variable, $\mathbf{a}_0$ a constant entity of the system A, and putting

$$\mathbf{a} = e^{\mathbf{R}t}\mathbf{a}_0,$$

$\mathbf{a}$ will be a function of t, which for t = 0 reduces to $\mathbf{a}_0$, and which satisfies the differential equation

$$\frac{d\mathbf{a}}{dt} = \mathbf{R}\mathbf{a}.$$

If two substitutions $\mathbf{R}$ and $\mathbf{S}$ are mutually commutative, that is $\mathbf{RS} = \mathbf{SR}$, the substitutions $\mathbf{R} + \mathbf{S}$, $\mathbf{RS}$, $\mathbf{R}^{-1}$, $e^{\mathbf{R}}$ are also commutative with them and among themselves, as are all the substitutions that one obtains from these by repeating the same operations. From the hypothesis of the commutability of **R** and **S**, one also has $e^{R+S} = e^{S+R}$.

The various substitutions that one can obtain from a substitution **R** and from numbers with the operations summation, multiplication, inversion, exponentiation, are mutually commutative.

79.   Let **x** be an entity of a linear system A, and let $\mathbf{y} = f(\mathbf{x})$ be a function of it, which also belongs to a linear system B. If **x′** is any entity of the system A, we will put as definition of the first member

$$\left(d\frac{\mathbf{x}'}{\mathbf{x}}\right) f(\mathbf{x}) = \lim \frac{1}{h}[f(\mathbf{x} + h\mathbf{x}') - f(\mathbf{x})],$$

the limit being attained by making the number h tend to zero. The entity $\left(d\frac{\mathbf{x}'}{\mathbf{x}}\right) f(\mathbf{x})$ belongs to the linear system B, and is a function of **x** and **x′**. To the entity **x′** one gives the name the differential of the independent variable **x**, and to the entity $\left(d\frac{\mathbf{x}'}{\mathbf{x}}\right) f(\mathbf{x})$ that of the differential of the function $f(\mathbf{x})$. When one has fixed the independent variable **x** and its differential **x′**, in place of $\left(d\frac{\mathbf{x}'}{\mathbf{x}}\right)$ we will write simply the letter d.

If e.g. $f(x)$ is a numerical function of the number x, having derivative $f'(x)$, one has

$$\left(d\frac{1}{x}\right) f(x) = f'(x), \qquad \left(d\frac{h}{x}\right) f(x) = f'(x)h.$$

If the linear system A has m dimensions, and that of B has n dimensions, upon putting

$$\mathbf{x} = x_1\mathbf{a}_1 + \ldots + x_m\mathbf{a}_m$$
$$\mathbf{x}' = x_1'\mathbf{a}_1 + \ldots + x_m'\mathbf{a}_m$$
$$\mathbf{y} = y_1\mathbf{b}_1 + \ldots + y_n\mathbf{b}_n,$$

where $\mathbf{a}_1 \ldots \mathbf{a}_m, \mathbf{b}_1 \ldots \mathbf{b}_n$ are the reference entities of the systems A and B, if the entity **y** is a function of **x**, the numerical variables $y_1 \ldots y_n$ are functions of the numbers $x_1 \ldots x_m$, and one has

$$\left(d\frac{\mathbf{x}'}{\mathbf{x}}\right)\mathbf{y} = \left(\frac{dy_1}{dx_1}x_1' + \ldots + \frac{dy_1}{dx_m}x_m'\right)\mathbf{b}_1$$
$$+ \ldots + \left(\frac{dy_n}{dx_1}x_1' + \ldots + \frac{dy_n}{dx_m}x_m'\right)\mathbf{b}_n$$

assuming the continuity of the partial derivatives of the y with respect to the x.

From the same hypothesis, supposing k is a real number, and **x** and **x'** are two entities of the linear system A, one has

$$\left(d\begin{smallmatrix}k\mathbf{x}'\\ \mathbf{x}\end{smallmatrix}\right) f(\mathbf{x}) = k\left(d\begin{smallmatrix}\mathbf{x}'\\ \mathbf{x}\end{smallmatrix}\right) f(\mathbf{x})$$

$$\left(d\begin{smallmatrix}\mathbf{x}'+\mathbf{x}''\\ \mathbf{x}\end{smallmatrix}\right) f(\mathbf{x}) = \left(d\begin{smallmatrix}\mathbf{x}'\\ \mathbf{x}\end{smallmatrix}\right) f(\mathbf{x}) + \left(d\begin{smallmatrix}\mathbf{x}''\\ \mathbf{x}\end{smallmatrix}\right) f(\mathbf{x}).$$

These last two formulas say that $\left(d\begin{smallmatrix}\mathbf{x}'\\ \mathbf{x}\end{smallmatrix}\right) f(\mathbf{x})$ is a distributive function of the entity **x'**.

The distributive operation or transformation, which when carried out on **x'**, gives for its result $\left(d\begin{smallmatrix}\mathbf{x}'\\ \mathbf{x}\end{smallmatrix}\right) f(\mathbf{x})$, one can indicate by $\left(d\begin{smallmatrix}*\\ \mathbf{x}\end{smallmatrix}\right) f(\mathbf{x})$, and one can call it the *derivative* of the entity **y** with respect to the entity **x**.

Applications

80.   1. Calling $a\alpha b$ any proposition whatever containing the two elements a and b of the same system, it may satisfy the logical equations

(1) $(a\alpha b) = (b\alpha a)$

(2) $(a\alpha b) \cap (b\alpha c) < (a\alpha c).$

Thus if by $a\alpha b$ one means the proposition

'the number a is the square of the number b',

then neither of the propositions (1) and (2) is satisfied. If however by the same symbol one means

'the number a is equal and opposite in sign to the number b'
'the right-line a intersects the right-line b'
'the right-line a is perpendicular to the right-line b',

(1) is satisfied, but not (2). The propositions

'the number a is greater than the number b'
'the number a is a multiple of the number b',

satisfy condition (2), but not (1). The propositions

'the number a is equal to the number b'
'the number a is congruent with the number b, with respect to a fixed modulus'
'the right-line a is parallel to the right-line b'
'the right-line a coincides with b'
'the figure a is superposable on the figure b', etc.

satisfy the two conditions (1) and (2), whence they can be considered equivalences (N. 72).

2. The *identity* of the two entities a and b, that is the affirmation that a and b are two names belonging to the same entity, is a particular equivalence. But between the entities of the same system one can establish several species of equivalence. Thus if one considers the rectilinear segments AA', BB', ... described by a point that travels from their origins to their terms, the identity of two segments is equivalent to the identity of their origins and their terms. But the propositions 'the segment AB is superposable on $A'B''$, 'the right-line AB is parallel to $A'B''$, 'the right-line AB coincides with $A'B''$, etc., can be assumed as alternative equivalences.

3. Every equivalence between the entities of a system, different from the identity, is equivalent to the identity between the entities that one obtains from those of the system, upon abstracting from one and all those properties that distinguish an entity from its equivalent. Thus the equivalence 'the segment AB can be superposed on $A'B''$ is equivalent to the identity between the entities one obtains from each segment upon abstracting all the properties that distinguish it from those with which it is superposable. The entity that results from this abstraction is called the *magnitude* of the segment; the preceding equivalence is therefore equivalent to the identity of the magnitudes of the two segments. If we agree to indicate the identity with the symbol $=$, the equivalence now considered can be written

$$mg \ AB = mg \ A'B'.$$

Analogously, the equivalence 'the right-line AB is parallel to $A'B''$ can be written

$$\text{direction } AB = \text{direction } A'B',$$

and so on.

4. The logical product (conjunction) of two equivalences is again an equivalence; not in general the negation of an equivalence, or the logical sum of two equivalences.

81.   1. Consider the systems of forces applied to a figure of invariant form, or variable according to determinate laws. Attributing to the symbols $=$, $+$, 0 the meaning of the mechanical equivalence of two systems, of the mechanical superposition of two systems, of a system in equilibrium, and for the product of a system by a number meaning the system that one obtains upon multiplying all the forces by that number, one deduces that the systems of forces are entities of a linear system. The number of dimensions of this linear system depends on the variability of the figure; if it is rigid and free in space, the system is of 6 dimensions.

2. Consider the complete algebraic function $f(x)$ of a numerical variable x. Meaning by $f_1(x) = f_2(x)$ the identity of the values of $f_1(x)$ and $f_2(x)$ whatever the value of x may be, by $f_1(x) + f_2(x)$ the complete function the sum of $f_1(x)$ and $f_2(x)$, by $mf(x)$, where m is a number, the product of the number m with the function $f(x)$, and by 0 a function zero for every value of x, the functions considered are entities of a linear system.

If one considers only the complete functions of degree n, these comprise a linear system of $n + 1$ dimensions; the complete functions of any degree whatever form a system of infinite dimension.

3. From hypotheses 1, 2, 3 and 4 made in N. 72 one deduces $(\mathbf{a} = \mathbf{b}) = (\mathbf{a+c} = \mathbf{b+c})$. Vice versa, from hypotheses 1, 2 and 3, and from that logical equivalence, one can deduce 4.

82.   1. If $u = \mathbf{R}a$ is a distributive numerical function of the formations of the first species A on a right-line, one can determine another formation of the first species P on the same right-line for which, whatever A may be, one has

$$RA = PA,$$

where, in the second member, by PA is meant the number $\frac{PA}{u}$ (Vide N. 33).

If $\mathbf{R} = \begin{pmatrix} u_1 & u_2 \\ A_1 & A_2 \end{pmatrix}$, that is if u_1 and u_2 are the values of u corresponding to the two formations A_1 and A_2, whose product is not to be zero, it suffices to make

$$P = \frac{1}{A_1 A_2}(u_2 A_1 - u_1 A_2).$$

2. If $u = \mathbf{R}A$ is a distributive numerical function of the formation of the first species A in a plane, one can determine a formation of the second species p in the plane for which, whatever A may be, one has

$$RA = pA,$$

where, in the second member, by pA is meant the number $\frac{pA}{\omega}$ (Vide N. 43).

3. If $u = \mathbf{R}a$ is a distributive numerical function of the formation of the second species a in the plane, one can determine a formation of the first species P in the plane for which, whatever a may be, one has

$$Ra = Pa.$$

4. If $u = \mathbf{R}U$ is a distributive numerical function of the vector U in space, one can determine a vector I for which, whatever U may be, one has

$$RU = I|U.$$

5. If $u_1 = \mathbf{R}_1 A$, $u_2 = \mathbf{R}_2 a$, $u_3 = \mathbf{R}_3 \alpha$ are distributive numerical functions of the formations A, a, α of first, second and third species in space, one can determine formations π, i, P of third, second and first species for which, whatever A, a, α may be, one has

$$\mathbf{R}_1 A = \pi A, \quad \mathbf{R}_2 a = pa, \quad \mathbf{R}_3 \alpha = P\alpha.$$

83. **1.** Let A, B be two formations of the first species on a right-line, A', B' two other formations of the first species on a second right-line.

Supposing AB is not zero, the transformation

$$\mathbf{T} = \begin{pmatrix} A' & B' \\ A & B \end{pmatrix}$$

is called a *homography*. This makes correspond to every formation of the first species on the first right-line a formation of the same species on the second; to A and B correspond A' and B'. If $A'B'$ is not zero, the transformation $\mathbf{T}$ permits an inversion, and $\mathbf{T}^{-1} = \begin{pmatrix} A & B \\ A' & B' \end{pmatrix}$ makes correspond to every formation of the second right-line a formation of the first.

2. If $P = xA + yB$ and $Q = x'A + y'B$ are formations of the first right-line, upon putting $\mathbf{T}P = P'$, $\mathbf{T}Q = Q'$, one has $P' = xA' + yB'$, $Q' = x'A' + y'B'$. One deduces $PQ = (xy' - x'y)AB$, $P'Q' = (xy' - x'y)A'B'$, whence

$$\frac{PQ}{AB} = \frac{P'Q'}{A'B'}.$$

One deduces from this that the double ratio (N. 36, 2) of four formations of the first right-line is equal to that of their correspondents.

3. If A, B, A', B' are simple points, the homography is called a *similitude*. In this case, to each simple point corresponds a simple point, to a vector, a vector.

4. If A, B, A', B' are four formations of the first species on the same right-line, the transformation $\mathbf{T}$ is a substitution.

Between the substitution $\mathbf{T}$ and its square $\mathbf{T}^2$ there holds the relation

$$\mathbf{T}^2 - p\mathbf{T} + q = 0,$$

where one has put

$$p = \frac{AB' + A'B}{AB}, \qquad q = \frac{A'B'}{AB}.$$

The numbers p and q are important in the study of the substitution $\mathbf{T}$. If q is zero, $\mathbf{T}$ does not admit an inversion.

If p is zero, the square of $\mathbf{T}$ equals the number $-q$.

5. A non-numerical substitution, whose square is a number, is called an *involution*. If **T** is an involution, $\mathbf{T}^2 P$ coincides with P.

6. The substitution

$$\mathbf{I} = \begin{pmatrix} A & -B \\ A & B \end{pmatrix}$$

has 1 for its square, and thus is an involution; the elements A and B coincide with their correspondents. Every substitution, other than 1 and -1, whose product is 1, is reducible to the preceding form.

The substitution $\begin{pmatrix} Q & P \\ P & Q \end{pmatrix}$, whose square equals 1, can be reduced to the preceding form, making $A = P + Q, B = P - Q$.

7. The substitution

$$\mathbf{I} = \begin{pmatrix} B & -A \\ A & B \end{pmatrix}$$

has for its square -1, and is therefore an involution; in it no element coincides with its correspondent. Every substitution whose square equals -1 is reducible to the preceding form in infinitely many ways, one being able to take at random the element A.

8. The substitution

$$\mathbf{I} = \begin{pmatrix} A & 0 \\ A & B \end{pmatrix}$$

has 0 for its square. Every nonzero substitution whose square is zero can be reduced to the preceding form; the element A can be taken at random, provided it does not coincide with B, which corresponds to zero.

9. A substitution **T**, if the numbers p and q satisfy the condition $\frac{p^2}{4} - q > 0$, can be reduced in a unique way to the form

$$\mathbf{T} = x + y\mathbf{I},$$

where x and y are real numbers and **I** is an involution, whose square equals 1 (Vide Exercise 6). By the same hypothesis, the substitution **T** can be reduced to the form

$$\mathbf{T} = \begin{pmatrix} k_1 A & k_2 B \\ A & B \end{pmatrix}$$

where k_1 and k_2 are the real roots of the equation $k^2 - pk + q = 0$.

10. A substitution **T**, where $\frac{p^2}{4} - q < 0$, is reducible in a unique way to the form

$$\mathbf{T} = x + y\mathbf{I},$$

where x and y are real numbers and **I** is a substitution whose square equals -1 (Vide Exercise 7).

11. A substitution **T**, where $\frac{p^2}{4} - q = 0$, is reducible to the form

$$\mathbf{T} = x + \mathbf{I},$$

where **I** is a substitution whose square equals 0 (Vide Exercise 8).

12. If $\mathbf{I}^2 = 1$, one has

$$e^{x+y\mathbf{I}} = e^x(\cosh y + \mathbf{I}\sinh y);$$

if however $\mathbf{I}^2 = -1$, one has

$$e^{x+y\mathbf{I}} = e^x(\cos y + \mathbf{I}\sin y);$$

and if $\mathbf{I}^2 = 0$, one has

$$e^{x+y\mathbf{I}} = e^x(1 + \mathbf{I}y).$$

13. If A, B are simple points, I a vector, the transformation

$$\begin{pmatrix} A+I & U \\ A & I \end{pmatrix}$$

represents a *translation*, which to each point P makes correspond the point $P + I$; the transformation

$$\begin{pmatrix} A & -I \\ A & I \end{pmatrix}$$

represents a *symmetry*, which to each point makes correspond that one symmetric with respect to the point A.

14. If one sets

$$\mathbf{I} = \begin{pmatrix} B & A \\ A & B \end{pmatrix}, \quad \mathbf{J} = \begin{pmatrix} A & -B \\ A & B \end{pmatrix}, \quad \mathbf{K} = \begin{pmatrix} B & -A \\ A & B \end{pmatrix}$$

one has

$$\mathbf{I}^2 = \mathbf{J}^2 = -\mathbf{K}^2 = 1; \quad \mathbf{IJ} = -\mathbf{JI} = \mathbf{K}; \quad \mathbf{IK} = -\mathbf{KI} = \mathbf{J}; \quad \mathbf{KJ} = -\mathbf{JK} = \mathbf{I}.$$

Every substitution **T** can be reduced to the form

$$\mathbf{T} = m + x\mathbf{I} + y\mathbf{J} + z\mathbf{K},$$

where m, x, y, z are numbers (coordinates of **T** with respect to the reference substitutions 1, **I, J, K**).

Preserving the notation of Exercise 4 in the present No., one has

$$p = 2m, \quad q = m^2 - x^2 - y^2 - z^2.$$

If m = 0, **T** is an involution.

15. If **S** and **T** are two involutions, the sum **S** + **T** is also an involution. The transformation **ST** + **TS** is equal to a number; if this number is zero, the two involutions are called *harmonic*, and one has **ST** = −**TS**.

16. The equation

$$T^n + a_1 T^{n-1} + \ldots + a_{n-1}T + a_n = 0,$$

where $a_1 \ldots a_n$ are given numbers, and **T** is a substitution to be determined of formations of the first species on a right-line, is satisfied

a) Upon setting **T** = x, where x is a (real) number which, substituted into the preceding equation in place of **T**, satisfies it.

b) If x_1 and x_2 are two unequal numbers which, substituted in place of **T**, satisfy the equation, one can set

$$T = \begin{pmatrix} x_1 A & x_2 B \\ A & B \end{pmatrix}$$

where A and B are arbitrary formations.

c) If the complex number $x + y\sqrt{-1}$ substituted in place of **T** satisfies the given equation, one can put **T** = x + y**I**, where **I** is any substitution whatever whose square equals −1 (Vide Exercise 7).

Every solution of the equation proposed is included in the preceding.[2]

84.   1. What was said in the preceding about substitutions of formations of the first species on a right-line is applicable to all the substitutions of entities of linear systems in two dimensions; as a particular case to the system of vectors in a plane.

The distributive operation ⊥, carried out on vectors in a plane, transforms the vectors into vectors; thus it is a substitution; and since (N. 40)

$$\perp^2 = -1,$$

it represents an involution. Consequently every expression of the form x + y ⊥, where x and y are (real) numbers represents a substitution of the vectors of the plane. If U is a vector, upon putting

$$V = (x + y \perp)U,$$

one has $V = xU + y \perp U$, whence V is a vector having for coordinates x and y with respect to the vectors U and $\perp U$.

2. One has, if x is a number:

$$e^{x\perp} = \cos x + \perp \sin x.$$

The expression

$$V = r\, e^{x\perp} U,$$

where U is a vector, x a real number, r a positive number, represents the vector that one obtains upon rotating the vector U by the angle x, and then multiplying it by r.

3. The equation

$$P_1 = O + e^{x\perp}(P - O),$$

where O is a simple point, makes correspond to every point P of the plane a point P_1 that one obtains upon rotating P about O by the angle x.

4. Upon putting

$$P_1 = P + I, \qquad P_2 = O + e^{x\perp}(P_1 - O),$$

one has that P_2 is the new position of the point P after a translation represented by the vector I, and a rotation about O by the angle x. One recovers

(1) $$P_2 = O + e^{x\perp}I + e^{x\perp}(P - O).$$

Let one determine the point C in such a way that

(2) $$C = O + e^{x\perp}I + e^{x\perp}(C - O).$$

One recovers

$$C = O + \frac{e^{x\perp}}{1 - e^{x\perp}}I = O + \frac{1}{e^{-x\perp} - 1}I = O + \frac{\perp}{2\sin\frac{1}{2}x}e^{(x/2)\perp}I.$$

Subtracting (2) from (1), one deduces

$$P = C + e^{x\perp}(P - C),$$

that is, a translation followed by a rotation is equivalent to a rotation alone.

5. A rotation by the angle x about a point O_1 followed by a rotation by the angle y about a point O_2 is equivalent to a single rotation by the angle $x + y$ about the point

$$O_1 + \frac{1 - e^{y\perp}}{1 - e^{(x+y)\perp}}(O_2 - O_1) = O_1 + \frac{\sin\frac{1}{2}y}{\sin\frac{1}{2}(x+y)}e^{-(x/2)\perp}(O_2 - O_1)$$

$$= O_2 + \frac{\sin\frac{1}{2}x}{\sin\frac{1}{2}(x+y)}e^{(y/2)\perp}(O_1 - O_2),$$

supposing $\sin\frac{1}{2}(x + y)$ is not zero; if this should occur, the same two rotations are equivalent to a translation.

6. If U and V are two vectors in the plane, the first of which is not zero, one can always determine, and in a unique way, a transformation of the form $x + y \perp$

in such a way that there results

$$V = (x + y \perp)U.$$

This transformation one can indicate by $\frac{V}{U}$. One deduces

$$\frac{V}{U} = \begin{pmatrix} V & \perp V \\ U & \perp U \end{pmatrix} = \frac{U \perp V}{U \perp U} + \frac{UV}{U \perp U} \perp .$$

7. The equation

$$\frac{V}{U} = \frac{V'}{U'},$$

says that $\frac{mg\, V}{mg\, U} = \frac{mg\, V'}{mg\, U'}$, and the angles (U, V) and (U', V') are equal.

8. The equation

$$\frac{C - A}{B - A} = \frac{C' - A'}{B' - A'}$$

where A, B, C, A', B', C' are points, says that the triangles ABC and $A'B'C'$ are similar and of the same sense.

One deduces a construction with the straight-edge and compass from the vectors $\frac{V}{U}W$, where U, V, W are given vectors.

9. Let A, B, A', B' be four points in the plane; one wishes to determine the point C for which the triangles CAB and $CA'B'$ are similar and of the same sense. The point C must satisfy the equation

$$\frac{C - A}{B - A} = \frac{C - A'}{B' - A'},$$

which, solved for C, gives

$$C = A + \frac{B - A}{(B - A) - (B' - A')}(A' - A),$$

and thus one can construct the point C, unless $B - A = B' - A'$.

10. The transformation $\begin{pmatrix} I & \perp I \\ I & \perp I \end{pmatrix}$, where I is a vector in the plane, makes correspond to each vector that *symmetric* with respect to I.

11. The transformations $\begin{pmatrix} I & -J \\ I & J \end{pmatrix}$ and $\begin{pmatrix} J & -I \\ I & J \end{pmatrix}$, where I and J are any vectors in the plane, are the most general transformations whose squares equal 1 and -1 respectively.

12. If I, J, I', J', U are 5 given vectors in the plane, the construction of the vector

$$\begin{pmatrix} I' & J' \\ I & J \end{pmatrix}U$$

is a linear problem (N. 15).

13. Setting

$$P = O + e^{tS}I,$$

where O is a point, I a vector, $\mathbf{S}$ a substitution of vectors in the plane, t a numerical variable, upon varying t the point P describes a curve in the plane. If $\mathbf{S} = 1$, it is a *circle*. If $\mathbf{S}$ is any substitution whatever whose square equals -1, one has an *ellipse*. If $\mathbf{S}$ is a non-numerical substitution whose square equals 1, the curve is an *hyperbola*; if $\mathbf{S} = \begin{pmatrix} I & \perp I \\ I & \perp I \end{pmatrix}$, one has the *equilateral* hyperbola. If $\mathbf{S} = \frac{V}{U}$, where U and V are any two vectors whatever, one has the *logarithmic spiral*. In each case, differentiating, one has

$$\frac{dP}{dt} = \mathbf{S}(P - O), \qquad \frac{d^2P}{dt^2} = \mathbf{S}^2(P - O),$$

and thus one can easily construct the first and second derivatives, and consequently the tangent and the center of the osculating circle of the curve.

85.   1. If $ABCA'B'C'$ are formations of the first species in a fixed plane, the substitution

$$\mathbf{T} = \begin{pmatrix} A' & B' & C' \\ A & B & C \end{pmatrix}$$

is called a *homography*. To each formation of the first species in the plane corresponds another formation of the first species. To collinear elements correspond collinear elements. If PQR are three formations in the plane, and $P'Q'R'$ are their correspondents, one has

$$\frac{P'Q'R'}{A'B'C'} = \frac{PQR}{ABC}.$$

2. If $A'B'C'$ is not zero, the transformation $\mathbf{T}$ permits inversion. But if $A'B'C' = 0$, A', B', C' are collinear, and one can determine three numbers m, n, p, not all zero, for which $mA' + nB' + pC' = 0$. Under this hypothesis, to each element of the plane corresponds an element collinear with $A'B'C'$, and to the nonzero elements $mA + nB + pC$ correspond zero. In this case the homography is called *special*.

3. One has, if x is any number whatever,

$$\mathbf{T} - \mathbf{x} = \begin{pmatrix} A' - xA & B' - xB & C' - xC \\ A & B & C \end{pmatrix},$$

and for the homography $\mathbf{T} - \mathbf{x}$ to be special, it is necessary that

$$(A' - xA)(B' - xB)(C' - xC) = 0,$$

or, developing,

$$x^3 - px^2 + qx - r = 0,$$

where one has put

$$p = \frac{A'BC + AB'C + ABC'}{ABC}, \quad q = \frac{A'B'C + A'BC' + AB'C'}{ABC}, \quad r = \frac{A'B'C'}{ABC}.$$

4. One has between the powers 0, 1, 2, 3 of **T** the relation

$$\mathbf{T}^3 - p\mathbf{T}^2 + q\mathbf{T} - r = 0.$$

5. If one supposes $ABC \; A'B'C'$ are simple points, the transformation **T** is called an *affinity*. The areas of the triangles formed by corresponding points are proportional. If in addition the figure $A'B'C'$ is similar or equal to ABC, **T** represents a *similitude* in the plane, or an *equivalence*.

6. If $ABCD \; A'B'C'D'$ are formations of the first species in space, the substitution

$$\mathbf{T} = \begin{pmatrix} A' & B' & C' & D' \\ A & B & C & D \end{pmatrix}$$

represents a *homography*, for which to each element corresponds another element, and to collinear or coplanar elements correspond collinear or coplanar elements. If the elements that are linked in **T** are simple points, **T** represents an *affinity*. The volumes of tetrahedra corresponding in an affinity are proportional. The affinity includes as particular cases *similitude* and *equivalence*.

86. **1.** In the plane, if O and P are formations of the first species, u the unit bivector, one indicates by $\perp (O^*.u)$ that distributive operation which, carried out on a point P, gives for its result $\perp (OP.u)$. Then the transformation

$$e^{x \perp (O^*.u)}$$

carried out on the formation of the first species P in the plane, represents, if O is a simple point, a rotation by the angle x about the point O; if O is a point with mass m, it represents a rotation about O by the angle mx; if O is a vector, it represents a translation.

2. In space, if I is a vector, in magnitude equal to unity, one indicates by $|(I^*)$ the distributive operation which, carried out on a vector U gives for its result $|(IU)$. Then the transformation

$$e^{x|(I^*)}$$

carried out on vectors in space, represents a rotation of these vectors by the angle x about the direction of I.

3. In space, if a is any formation of the second species, ω the unit trivector, one indicates by $|(a * .\omega)$ that distributive operation which, carried out on a formation of the first species P gives for its result $|(aP.\omega)$. Then the transformation

$$e^{X|(a^* .\omega)}$$

carried out on the formations of the first species in space represents either a rotation about a right-line, or a translation, or a rotation and a translation together, according as a is reducible to a line, or a bivector, or to a line and a bivector together.

87. The rules for finding the differentials of entities that are functions of other entities (supposing that they are variable entities, and those functions belong to a linear system) are analogous to those that serve for entities that are functions of numbers. One has:

1. If the entity $\mathbf{u}$ does not depend on $\mathbf{x}$, one has

$$\left(d\frac{\mathbf{y}}{\mathbf{x}} \right) \mathbf{u} = 0.$$

2. If $\mathbf{Rx}$ is a distributive function of $\mathbf{x}$ one has

$$\left(d\frac{\mathbf{y}}{\mathbf{x}} \right) \mathbf{Rx} = \mathbf{Ry}.$$

3. $d(\mathbf{u} + \mathbf{v}) = d\mathbf{u} + d\mathbf{v}.$

4. If $\mathbf{uv}$ indicates a distributive function of the entities $\mathbf{u}$ and $\mathbf{v}$, such as the product of two numbers, the product of a number by a geometric formation, or by an entity of a linear system, the progressive or regressive product of two geometric formations, the product of a transformation by an entity of a linear system, the product of two transformations, supposing that $\mathbf{u}$ and $\mathbf{v}$ depend on the entity $\mathbf{x}$, one always has

$$d.\mathbf{uv} = d\mathbf{u}.\mathbf{v} + \mathbf{u}.d\mathbf{v}.$$

5. Thus, if $\mathbf{S}$ is a substitution of the entities of a linear system, one has

$$d.\mathbf{S}^2 = \mathbf{S}.d\mathbf{S} + d\mathbf{S}.\mathbf{S},$$
$$d.\mathbf{S}^3 = \mathbf{S}^2.d\mathbf{S} + \mathbf{S}.d\mathbf{S}.\mathbf{S} + d\mathbf{S}.\mathbf{S}^2, \text{ etc.}$$
$$d.\mathbf{S}^{-1} = -\mathbf{S}^{-1}.d\mathbf{S}.\mathbf{S}^{-1}$$
$$d.\mathbf{S}^{-2} = -\mathbf{S}^{-1}.d\mathbf{S}.\mathbf{S}^{-2} - \mathbf{S}^{-2}.d\mathbf{S}.\mathbf{S}^{-1}, \text{ etc.}$$

6. If $\mathbf{S} = \begin{pmatrix} \mathbf{b}_1 & \cdots & \mathbf{b}_n \\ \mathbf{a}_1 & \cdots & \mathbf{a}_n \end{pmatrix}$ is any substitution whatever of the entities of a linear system, and if $\mathbf{a}_1 \ldots \mathbf{a}_n \ \mathbf{b}_1 \ldots \mathbf{b}_n$ are entities that are functions of the entity $\mathbf{x}$, one has

$$dS = \begin{pmatrix} db_1 & \cdots & db_n \\ a_1 & \cdots & a_n \end{pmatrix} - \begin{pmatrix} b_1 & \cdots & b_n \\ a_1 & \cdots & a_n \end{pmatrix} \begin{pmatrix} da_1 & \cdots & da_n \\ a_1 & \cdots & a_n \end{pmatrix}.$$

88.   1. If $f(\mathbf{x})$ is an entity that is a homogeneous function of the entity $\mathbf{x}$, of degree n, that is, is such that, whatever the number t, one has $f(t\mathbf{x}) = t^n f(\mathbf{x})$, one has

$$\left(d\frac{\mathbf{x}}{\mathbf{x}} \right) f(\mathbf{x}) = n f(\mathbf{x}).$$

2. If $f(\mathbf{x})$ is a homogeneous function of degree 0 in $\mathbf{x}$, one has

$$\left(d\frac{\mathbf{x}}{\mathbf{x}} \right) f(\mathbf{x}) = 0.$$

3. If $f(\mathbf{x})$ is a complete homogeneous function of degree n if $\mathbf{x}$ (that is, the coordinates of $f(\mathbf{x})$ are complete homogeneous functions of degree n of the coordinates of $\mathbf{x}$), letting p and q be numbers, and $\mathbf{y}$ an entity of the same linear system to which $\mathbf{x}$ belongs, one has

$$f(p\mathbf{x} + q\mathbf{y}) = p^n f(\mathbf{x}) + \frac{p^{n-1}q}{1}\left(d\frac{\mathbf{y}}{\mathbf{x}} \right) f(\mathbf{x}) + \frac{p^{n-2}q^2}{2!}\left(d\frac{\mathbf{y}}{\mathbf{x}} \right)^2 f(\mathbf{x})$$
$$+ \ldots + \frac{q^n}{n!}\left(d\frac{\mathbf{y}}{\mathbf{x}} \right)^n f(\mathbf{x}).$$

The successive differentials $\left(d\frac{\mathbf{y}}{\mathbf{x}} \right) f(\mathbf{x})$, $\left(d\frac{\mathbf{y}}{\mathbf{x}} \right)^2 f(\mathbf{x})$, ... are also called the *polar functions* of successive orders obtained from $f(\mathbf{x})$ with the operation $\left(d\frac{\mathbf{y}}{\mathbf{x}} \right)$. They are complete homogeneous functions of degree $n-1$, $n-2$, ... in $\mathbf{x}$, and of degree 1, 2, ... in $\mathbf{y}$.

4. If $u = f(X)$ is a complete homogeneous numerical function of the formation X of the first species in the plane, then $\left(d\frac{Y}{X} \right) u$ is a distributive function of Y, which is another formation of the first species in the plane. Thus one can determine a formation p of the second species in the plane, dependent on X, but not on Y, such that one has $\left(d\frac{Y}{X} \right) u = pY$. The formation p one can indicate by $\left(d\frac{*}{X} \right) u$, and one can call it the *polar formation* of X with respect to $f(X)$.

If X is a point of the locus of the equation $f(X) = 0$, one has $pX = 0$; thus the formation $p = \left(d\frac{*}{X} \right) u$ is a line passing through the point X; it is the tangent in that point to the curve of the equation $f(X) = 0$.

5. If $u = f(x)$ is a complete homogeneous numerical function of the formation x of the second species in the plane, then $\left(\mathrm{d}\dfrac{y}{x}\right) u$ is a distributive numerical function of the formation of the second species y. One can therefore determine a formation of the first species P that one can also indicate by $\left(\mathrm{d}\overset{*}{\underset{x}{}}\right) u$ such that, whatever y may be, one has $\left(\mathrm{d}\dfrac{y}{x}\right) u = Py$. The formation P can be called the *pole* of x. If x is a line that satisfies the equation $f(x) = 0$, and P is a point, one has $Px = 0$, whence P lies on x; the point P is the point of contact of the right-line x with the envelope of the equation $f(x) = 0$.

6. What has been said can easily be extended to formations of the first, second, and third species in space.

89.   **1.** In a fixed plane let X be a variable formation of the first species, $a_0 a_1 \ldots a_n$ be given formations of the second species, and let $m_0 m_1 \ldots m_n$ be given numbers: One puts

$$f(X) = m_0 (a_0 X)^2 + m_1 (a_1 X)^2 + \ldots + m_n (a_n X)^2 = \sum_i m_i (a_i X)^2.$$

One has that $f(X)$ is a numerical function of second degree in X. One has

$$\tfrac{1}{2}\left(\mathrm{d}\frac{Y}{X}\right) f(X) = \sum m_i a_i X . a_i Y,$$

whence the polar of X is

$$\tfrac{1}{2}\left(\mathrm{d}\overset{*}{\underset{X}{}}\right) f(X) = \sum m_i a_i X . a_i,$$

2. The intersection, or regressive product, of the two formations of the second species $\tfrac{1}{2}\left(\mathrm{d}\overset{*}{\underset{X}{}}\right) f(X)$ and $\tfrac{1}{2}\left(\mathrm{d}\overset{*}{\underset{Y}{}}\right) f(Y)$ can be put into the form

$$\sum m_i m_j (a_i a_j . XY) a_i a_j,$$

or, putting $XY = x$, into the form

$$\sum m_i m_j a_i a_j x . a_i a_j,$$

which is a formation of the first species. It is called the *pole* of x.

3. Dependent on $f(X)$ is the function of second degree in x

$$F(x) = \sum m_i m_j (a_i a_j x)^2$$

and the constant

$$A = \sum m_i m_j m_k (a_i a_j a_k)^2.$$

The equation $F(X) = 0$ represents a locus of second order. The formation of the second species (right-line) x contains no element, or one or two, that satisfy the equation, according as $F(X) \gtreqless 0$. This locus contains no vector, or one, or two (that is, it is an ellipse, or a parabola or an hyperbola) according as $F(X) \gtreqless 0$.

Through the formation of the first species (point) X passes no, or one, or two formations (right-line) x that satisfy $F(x) = 0$, according as $Af(X) \gtreqless 0$.

4. If in the function $f(X)$ one supposes $a_1 \ldots a_n$ are lines in magnitude equal to the unit of measure, the numbers $m_1 \ldots m_n$ are equal to unity, a_0 is the unit bivector u, and m_0 any number $-k$, the function becomes

$$f(X) = (a_1 X)^2 + \ldots + (a_n X)^2 - k(uX)^2,$$

and the equation $f(X) = 0$ is satisfied by all the points X for which the sum of the squares of the distances from the right-lines $a_1 \ldots a_n$ is a given quantity k.

5. If X is a vector, the polar of X will be, upon observing that $uX = 0$,

$$\tfrac{1}{2} \left(d \overset{*}{X} \right) f(X) = a_1 X.a_1 + \ldots + a_n X.a_n.$$

This right-line is called the *conjugate diameter* of the direction of the vector X. To construct it, one constructs the lines $a_i X.a_i$, that is the lines that lie on the right-lines given, and are proportional to the sines of the angles that they make with the direction X; their sum will be a line lying on the diameter sought.

6. The conjugate diameters of the various directions X pass through a fixed point, called the *center*, which is the pole of the bivector u (right-line at infinity). This, by virtue of formula 2, can be written

$$\sum_{ij} a_i a_j u.a_i a_j,$$

and thus it is the barycenter of the points of intersection (at a finite distance) of the right-lines considered, with masses equal to the squares of the sines of the angles that they make.

This center is also the point for which the sum of the squares of the distances from the given right-line is a minimum.

7. The directions of the axes, that is of the conjugate diameters to vectors perpendicular to them, can be obtained by the following construction. Taking a vector X at random, one determines the vectors $X_1 \ldots X_n$ symmetric to X with respect to $a_1 \ldots a_n$; the directions that divide in half the angle of X with $X_1 + \ldots + X_n$ are those of the axes.

90.   1. Let u be a numerical function of the position of a simple point P in space. Since the difference of two points is a vector, one can take for the differential of the point P a vector I. Then $\left(d \overset{*}{\underset{P}{}} \right)$ u represents a number, a distributive function of the vector I. Thus one can determine a vector U, dependent on P, but not on I, such that, whatever I may be, one has

$$\left(d \overset{*}{\underset{P}{}} \right) u = U | I.$$

The vector U is called the *differential parameter* (of first order) of u.

If u represents the distance of the point P from a fixed point, or from a fixed right-line, or from a fixed plane, the differential parameter is a vector directed from the point to the fixed point, or normal to the fixed right-line or plane, in the sense that it goes from the fixed point or right-line or plane to the variable point, and in magnitude is equal to the unit of measure.

If $u = f(u_1, u_2, \dots, u_n)$ is a numerical function of the numbers $u_1 \dots u_n$ which are, in turn, functions of the point P, and if $U_1 \dots U_n$ are the differential parameters of $u_1 \dots u_n$, the differential parameter of u is $\frac{df}{du_1} U_1 + \dots + \frac{df}{du_n} U_n$.

The differential parameter U of u is normal to the surface locus of points for which u has the same value.

If for a point P u becomes a maximum or minimum, its differential parameter, if it exists, is zero (Vide *App. Geom.*, Chapter IV).

2. Let u be a numerical function of the position of a right-line in a fixed plane. Taking any line p, it determines the right-line p, and thus a value of u; and since the right-line tp, where t is a number, coincides with p, and since the value of u remains the same, one deduces that u is a homogeneous function of degree zero of each line p of the plane. One determines the formation of the first species (point or vector) $P = \left(d \overset{*}{\underset{p}{}} \right) u$. One has $Pp = 0$, that is P lies on the line p. If P is a point, it will be the point of contact of the right-line p with its envelope; if for a certain position of p the function u becomes a maximum or minimum, the formation of the first species $\left(d \overset{*}{\underset{p}{}} \right)$ u must be zero.

If $u = f(r_1, r_2, \dots, r_n)$, where $r_1 \dots r_n$ represent the distances of the fixed points $A_1 \dots A_n$ from the right-line p, calling $B_1 \dots B_n$ the bases of the perpendiculars dropped from the given points to p, and supposing $mg\ p = 1$, one has

$$\left(d \overset{*}{\underset{p}{}} \right) u = \frac{df}{dr_1} B_1 + \frac{df}{dr_2} B_2 + \dots + \frac{df}{dr_n} B_n.$$

(Vide *App. Geom.*, P. 329).

3. The same remarks can be repeated for a number that is a function of the position of a plane in space (Vide *App. Geom.*, P. 333).

4. If u is a numerical function of the position of a right-line in space, upon taking any line p, it determines a right-line, and thus a value of u; thus u is a function of every line in space, and is a homogeneous function of degree zero. Putting $q = \left(d \overset{*}{\underset{p}{}} \right) u$, q will be a formation of the second species in space that satisfies the condition $pq = 0$. Taking a point P on the right-line p, the plane Pq is the tangent along the right-line p of the cone locus of the right-lines passing through P, and for which u has a constant value. Producing through p a plane π, the point πq is the point of contact of p with the envelope of the right-lines contained in the plane π, and for which u has a constant value.

Editorial Notes

I am greatly obliged to Hongbo Li for providing these two Editorial Notes.

1. (p. 75) The reduction claimed cannot in fact be carried out. Thus, let

$$f = (B + C) \tan A + (C + A) \tan B + (A + B) \tan C,$$
$$g = A \tan 2A + B \tan 2B + C \tan 2C.$$

Using $A + B + C = \pi$, we can eliminate the angle C from f and g, and we have:

$$
\begin{aligned}
f = &- A \tan A(1 + \tan^2 B)/(1 - \tan A \tan B) \\
&- B \tan B(1 + \tan^2 A)/(1 - \tan A \tan B) \\
&+ C(\tan A + \tan B),
\end{aligned}
$$

$$
\begin{aligned}
g = &2A \tan A/(1 - \tan^2 A) \\
&+ 2B \tan B/(1 - \tan^2 B) \\
&- 2C(\tan A + \tan B)(1 - \tan A \tan B)/(1 - \tan^2 A - \tan^2 B \\
&+ \tan^2 A \tan^2 B - 4 \tan A \tan B).
\end{aligned}
$$

If f and g represent the same point, they must be equal up to a nonzero scalar factor. For this the coefficients of A, B, C in f and g must have the same ratio, i.e.,

$$
\begin{aligned}
&(1 + \tan^2 B)(1 - \tan^2 A)/(1 - \tan A \tan B) \\
=&(1 + \tan^2 A)(1 - \tan^2 B)/(1 - \tan A \tan B) \\
=&(\tan A + \tan B)(1 - \tan^2 A - \tan^2 B + \tan^2 A \tan^2 B \\
&- 4 \tan A \tan B)/(1 - \tan A \tan B).
\end{aligned}
$$

The first equality gives $\tan^2 A = \tan^2 B$, a constraint not satisfied by a generic triangle.

2. (p. 134) This statement is incorrect. Thus, let $f(x) = x^n + a_1 x^{(n-1)} + \ldots + a_n$. Peano did not consider the case when $f(x)$ has multiple factors. Let x_j be a root of f whose multiplicity is at least two. When x_j is real, then $T = x_j + T_0$, where T_0 is any nilpotent substitution, will satisfy $f(T) = 0$. When x_j is complex we can construct other solutions. It seems that these solutions, together with the a), b), c) proposed by Peano, make up all possible solutions.

Subject Index

Subject Index